The Philosophy of Practical Affairs

PHILOSOPHICAL PRACTICE

Series Editor: Lydia Amir, Tufts University

The last decades have witnessed a renewed interest in the power of philosophy to address everyday problems, on both an individual and a social scale. The outcome has been a theoretical and practical field called "philosophical practice," an original approach that highlights the timely and perennial need for philosophy. This series aims to bring to the academic public the best reflections that bear on the relation of philosophy to everyday life and to the contemporary world, as grounded in experience or arguments or both. It honors the founders of this innovative field while calling for new ways of empowering philosophy by demonstrating its relevance to individual and social concerns both inside and outside academia. It thus hopes to strengthen philosophy by bringing its potency to the attention of philosophers and scholars from other disciplines, as well as to students and the general public.

Titles in the Series

The Philosophy of Practical Affairs: An Introduction, by Joseph Agassi

The Philosophy of Practical Affairs

An Introduction

Joseph Agassi

LEXINGTON BOOKS
Lanham • Boulder • New York • London

Published by Lexington Books
An imprint of The Rowman & Littlefield Publishing Group, Inc.
4501 Forbes Boulevard, Suite 200, Lanham, Maryland 20706
www.rowman.com

86-90 Paul Street, London EC2A 4NE

British Library Cataloguing in Publication Information Available

Library of Congress Cataloging-in-Publication Data Available

ISBN 9781793651730 (cloth) | ISBN 9781793651747 (ebook)

Contents

Preface

This book is the response to the kind invitation of Lydia Amir. "If I am not mistaken," she wrote to me, "you have never written a book on the practical nature of philosophy, despite your having been active in it all your life." In a sense, this is obviously true: my very writing of this book testifies to it. In another sense, as she has noticed in her letter, this is obviously false: almost all I ever wrote is about it. For most of my career as a philosopher, the leading trends of philosophy were the advocacy of philosophy as unreason and the advocacy of reason as science, so that philosophy is nothing but the advocacy of unreason, perhaps with the proviso that science is kindly allowed to be a manifestation of reason. My own view is—as it always was—traditional: philosophy is nothing, I wrote, if it is not the advocacy of reason. This made me view philosophical research as problem-oriented. Usually, most of the problems that interest me are large-scale, mostly relating to science or to politics, more specifically, mostly relating to the interaction between science and politics—in any case, hardly ever directly addressing problems of everyday life. The reason is that problems of everyday life belong to a very different field of study, the one known as the philosophy of life, as opposed to what is famous as philosophy proper, namely, academic philosophy, the set of problems studied in traditional courses of philosophy as traditionally taught there.

This book presents attempts to take up the challenge of Lydia Amir and discuss the problems of everyday life, problems that from time to time, here and there, fate drops on normal twenty-first century Western individuals, busy with attempts to solve the practical problems that fate has thrown on our road. Thus, the intent behind my discussions concerns the usefulness of the treatment of daily problems within academic philosophy. It differs from most philosophy-of-life literatures, namely, from pop-philosophy—especially religious homilies or wisdom literature, including the (pseudo-)cabbalist or (pseudo-)Buddhist or otherwise (pseudo-)oriental, and their likes. For, these often ignore real, disturbing situations.

This book presents then the obvious question, traditionally worded in diverse ways, such as the following. What is a good life? What makes a life good or satisfactory? What is the (proper) aim of life? And so on. The question often appears under the name "the meaning of life" or its cognates. Pop-philosophy ignores this question or deals with it in a piecemeal fashion or uses it as an excuse for delving into popular theology, especially theodicy. The wording of the distinction of discourses as between pop-philosophy and philosophy proper sounds too pretentious; so let me word it as the distinction of discourses as between the philosophy of life and academic philosophy. Within academic philosophy, a discussion of this question and of the set of traditional answers to it concern what is just or unjust and wise or unwise—with a tacit or explicit accent on some version or variant of anti-intellectualism[1] or its opposite, often also within universalism (that addresses all human cultures) and particularism (that addresses my own culture). Each of these philosophies then needs discussion—of its strength and weakness—always together with indications of its impact on the practical question of the meaning of life. Indeed, many discourses on the meaning of life concern the question, what philosophy fits my political or socio-political dispositions best? The philosophy in question often conflicts with standard Western liberal parliamentary philosophy, yet this is seldom noted.

All this should help advance the discussion; what this advancement is also is not self-evident, so that it invites discussion too. At this stage, there is no escape from a discussion of the riddle of life: does life have meaning in the first place? What kind of answer to this question should we take sufficiently seriously to put on our agenda?[2] Any possible discussion of the second question above one may consider a philosophy: what kind of answer to the question of the meaning of life deserves attention? To this question, every culture has its own answer, its own philosophy. The study of any such philosophy usually belongs to the study of the culture that articulates its answer. The study of all of them taken together has a name too: it is the philosophy of life. The philosophy of life is then in contrast with academic philosophy or traditional philosophy or philosophy as an intellectual discipline—at times academics call it philosophy proper.[3]

This book then is intentionally on the borderline between philosophy proper and philosophy of life. It belongs to both. We can sum this up as the philosophy of Socrates, the grand master of all philosophy: rightly comprehended, students of philosophy proper and students of the philosophy of life should try to unite them as best they can and look forward to complete success of this process of unification.

My gratitude to Nimrod Bar-Am and Chen Yehezkeli for many corrections and improvements of the final draft, and to Quinn Williamson for his work on the bibliography.

NOTES

1. Anti-intellectualism is more common than it seems, since many anti-intellectualists veil their anti-intellectualism.

2. This raises the question, who determines "our" agenda? Fortunately, the answer here is obvious: it is both author and reader. When readers disagree, they can skip passages of a text or ignore it altogether. Unfortunately, schools may force readers to read texts that they do not care for. This misconduct will be ignored here, as this text will hardly ever serve the standard curriculum. See also next note.

3. The standard style of an academic text prescribes annotation, although all old texts and some important ones have none. Scholars do or do not add notes to their texts. Clearly, at times notes serve to browbeat readers; and this is appalling. (For more see my *Towards a Rational Philosophical Anthropology,* 1977, Introduction: Against the elitism of Excessive Scholarship.) At times, however, they come to add items that cannot integrate well in the text yet they may help some interested readers. At other times they answer some objections that may trouble the specialists among its intended readers but that are lost on most readers.

PART I

Background

Chapter 1

Universalism

The Siblinghood of Humanity

Most social scientists agree: one of the things that may make for any social group is the awareness of its members of their belonging to it. Yet awareness is problematic. To begin with, obviously, some groups, even some important groups, have nothing to do with awareness, such as blood groups, especially some odd ones, such as Rh negative.[1] How different this is from groups that thrive on members pride in their belonging to it, in their sense of comradeship with each other. At times, one acquires this kind of membership by immense efforts and dedication to a given cause—such as the group of decorated veterans and the group of Nobel Prize winners. At times membership rests on nothing more than joining a given group, such as the group of fans of a sport club, where the membership amounts to almost nothing more than the awareness of belonging to it and the passive participation in some of the major events of that club's activities. Since this awareness of belonging to the alliance is all that it takes, and since this is a matter of sheer decision, switching such alliances is very easy yet almost unthinkable—just because it is so easy. How different this is from the membership of supporters of communism. This demanded of all members of that group some sacrifice, and they make this sacrifice willingly. Yet eventually they learned about the brutal conduct of the Soviet leadership, and then, most of them left the group of supporters of communism with deep sadness. Others clung to their positions and declared the unpleasant stories about them lies or exaggerations or deviations.

The most basic group is the family, nuclear and extended,[2] and it enjoys the status of a natural phenomenon and this makes it relatively unshakeable.[3] Yet it is the sense of belonging to it that gives it significance: some people do not know who their cousins are. Yet they can still be proud members of their extended families. The tartans—the diverse patterns of the kilt that designate diverse clans of the Scottish highlanders—symbolize it well. For, it has lost its initial significance long ago. Nevertheless, it still speaks to different people in diverse ways. Classifying people in other ways, for example when we speak of all those with some specific genetic immunity, comes with the realization that they are not a club or a family, since most of them are unaware of their possession of that quality, much less of their belonging to the group defined by it. There are moving stories—real and fictitious—of the return of lost members of some family to the fold, regardless of whether the fold in question is nuclear or extended or more artificial.

Discussions of social groups, that is, of groups other than of blood relations, normally concern groups whose members are well aware of their belonging; they concern groups of people who are proud of it. During World War I, the military of the United States of America expanded immensely, leading to the creation of new divisions. No matter how one divides a group, it is always likely that some of its members remain unclassified. They are usually called "the rest."[4] Those who classified the soldiers created a new unit of the rest, which they called the Rainbow Division.[5] Sociologists checked attitudes of different members of the Rainbow Division; they found it surprising that members of that arbitrarily selected unit were proud to belong to it. Their pride meant that they considered themselves as having passed some selection procedure that distinguished them positively, whereas they were just members of the national guard that were left over in a selection process that they were not aware of. The awareness of belonging to a social group of this type is usually quite innocuous: it matters little what reason one provides for thinking well of oneself. What matters is just this: it is good to think of oneself positively: it assists one's self-esteem, and self-esteem is very useful for the development of autonomy, and autonomy is morally most valuable. This may find its expression in the concern of leaderships with the morale of their crowds.

It is on the matter of autonomy that science and religion clash most significantly—not in principle but in sheer historical fact: religions recommend and encourage thinking too highly of one's group and too poorly of oneself.[6] This recommendation for thinking poorly of oneself rests on some false hypothesis according to which the increase of thinking poorly of oneself raises one's social worth than thinking highly of oneself. The traditional reason for this is simple and rather primitive. All traditions rest on the claim that there is a permanent, unresolvable conflict between society and the individual: the individual is lazy, selfish and prone to commit infractions. This is the famous doctrine of the Beast in Man or of the Original Sin: as society demands dedication, it has to develop means for curbing human beastly dispositions. Thus, there is a constant need to coerce people to work against their inclination to evade their duties. This conclusion is famous; it is *the* work ethic. The right comment on this doctrine is simple: this doctrine has an overwhelming number of refutations as any doctrine can. Were it true, then all societies would be unstable in the extreme. Modern experience shows that liberal societies are more stable than traditional ones. This refutes all traditional views on human society, especially the idea that we need to impose the work ethic by the development of a sense of guilt. Only recently, efforts appeared to replace the work ethic with improving the quality of working life.

Liberalism is the most successful idea or philosophy or movement. Any way we look at it, we may easily note that the liberal part of the globe and the rich part of the world coincide.[7] Nevertheless, many social thinkers claim that it is a failure. This disparity may be due to the hostility to liberalism the (intellectualist) elitism breeds.

The illiberal work ethic is also obviously false. It is not only redundant; it is extremely harmful. It destroys the natural pleasure in work, and thus even the joy of life; it destroys the possibility of independent thinking and it is thus conservative in principle and thus it opposes all social progress. Since technological progress facilitates social progress, the work ethic tacitly opposes technological progress. To see this at work one may notice the very response to every invention of a new robot. Such an invention should please us, as it somehow reduces the need for human unpleasant toil. Instead, the response is the fear that it raises social problems by creating unemployment. This is easy to overcome by legislation to the effect that any individual who suffers because of progress should receive adequate compensation from the state.

In the United States of America, coal mining declined for obvious reasons. President Donald Trump, who won elections by the use of the slogan "Make America Great Again" meant by it, among other things, to reopen coalmines. This would be impossible had the state adequately compensated redundant coalminers. However, many experts deem impossible any compensation that amounts to full coverage of all loss—on the refuted allegation that it encourages laziness.

Compare medieval and modern societies in this respect. The doctrine of original sin was widespread in the stagnant mediaeval society. Modern science rested on the opposite idea: human nature is good. Both ideas are obviously questionable. Obviously, we all have mixed dispositions, some good some evil, most a mix of both. The more operatively significant idea was liberalism: its originator Niccolò Machiavelli has suggested that legislators should try to design the law such that it will be in the interest of citizens to abide by it. His reason was simple and scarcely questionable: even in oversight of goodwill, the selfishness of most people is more reliable than their fear of punishment. The important aspect of his proposal, however, is its tacit reliance on toleration: when tending one's interest one tends to ignore the vices of one's neighbor. This, however, valuable it is, does not suffice. Parliamentary democracy is impossible without the explicit endorsement of toleration as a principle: unless we tolerate disagreement, even on matters of religion, democracy is impossible. It is not clear why. After all, Athenian democracy held only between citizens, to the exclusion of women and children, of slaves, and of foreign residents. Toleration was never nearly sufficiently the rule, yet it is what has allowed for the drive for modernism—the drive that made the current scientific era possible.

Modernism is an idea that rose to prominence by middle of the seventeenth century. It was primarily the idea that research into the behavior of Mother Nature, the contribution of what we call the natural sciences, takes priority over the study and the dissemination of the classics, as well as over political activity. Natural science made it apparent that the long-standing traditional distinction of class is not natural: nobility is not a blood characteristic but a social one. This very simple idea led to the demand for equality—egalitarianism—that has led to the French Revolution. A failure though this revolution certainly was (as it was taken over by the Terror and then by the Napoleonic Wars), its egalitarian ideas soon became generally endorsed. Progressive thinkers still view the French Revolution with admiration, and they have a reason. Yet, to repeat, it was a failure, as it ended with the reign of terror and the Napoleonic Wars. This failure brought about a profound change. This change occurred in the early nineteenth century and contributed significantly to the rise of the forces responsible for World War I, World War II, and the Holocaust—the greatest catastrophe that befell humanity. This catastrophe has revived he doctrine of the Beast in Man or of original sin in a modified form that has linked it to a new path to salvation: the individual is worthless but can find redemption by subordinately seeking pride in belonging to the newly discovered group that counts, namely, the Nation.

There is some truth in this idea, stupid though it is. For, a nation is, initially, an extended family. One can integrate into a nation thoughtlessly, by being born. That is all it takes to belong to a family and yet the feeling of belonging (including its associated emotions) is very strong. This small truth is easy to exaggerate.[8] There are many strong arguments that show nationalism and national sentiment to be pernicious rather than benign. To begin with, the awareness of belonging to one's nation is of marginal importance. It receives its verbal expression by different references to other people as belonging or not belonging to a group: we feel differently toward people who belong to our group and people who do not—whatever our group and whatever we think it happens to be. We speak about us as different from them. This awareness goes with the attribution of difference in characteristics between them and us, whatever we take the difference to be; chief among them happen to be family ties and shared religion, but the feeling is of belonging, no matter to what. In this context, religion is not only the prescribed rituals; it is the religious culture or the culture in a more general sense; specifically, it is ways of doing things—*our* way of doing things. This we view generally as the distinction between our ways and theirs. Of course, our ways are right and their ways are not. After all, we are superior to them. In the rare situations in which we do not feel this way, we try to convert to their religion or to join their nation. Sharing their family is not as easy as converting to their religion; rather, we can abolish family superiority, or what we call nobility or skin color or whatever we take as the characteristic of the superiority in question. The abolition of inequality was the great attraction of the French Revolution. Yet, to date, French people of noble descent still refer to themselves with the titles of nobility.[9] They use this inequality to feel at home in their society. Common people there allow for this. Hence, it is feeling at home in some group that matters, not its causes.

The more sophisticated among us learn to avoid looking at the world while dividing it to the extremes that we often label black and white: after learning the ways of our neighbors in some detail, we find it wise to admit: we are not all white and our neighbors are not all black. Hence, although we still are superior to them, in some minor ways they may be superior to us. In this case, it will be wise of us to imitate them. This is not always simple and straightforward: it takes some *finesse* and caution, and we should do so gradually, so as not to disturb our feeling that we belong to an existing stable order. In some cases, our leaders forbid emulating our neighbors despite the admitted superiority of their ways. In some cases, the changes, though merely technical, cause change in the social order; this causes social instability and thus it causes trouble and thus to be avoided at (almost) all cost. Even the copying of a technique can be problematic, since the imported technique ousts some indigenous technique, thus competing with a technique that it replaces and thus damaging people who deploy it. This, incidentally, is one reason why poor countries, whose inhabitants dearly wish to acquire wealth, fail in this assignment: they refuse to let go of some cultural qualities that look unrelated to wealth, but deceptively so.[10]

The obstacle is obvious: the feel of belonging to a tribal group blocks the feel of belonging to a civil society. The link between the sense of belonging to a tribal group and poverty is not obvious. And yet the feel of belonging to a society is essential to belonging to civil society—as opposed to religious society or community—and civil society is the backbone of modern society and thus to wealth. This is why all reactionaries discuss the advantage of community over civil society and the burden of alienation, of the absence of the feeling of belonging. The reactionaries are right in stressing that alienation is painful. This then invites some remedy. The reactionaries suggest that the cure is some return to the community. This is an impossible dream. Bigots even oppose the senses of well-being and of self-worth that many modernists advocate.[11]

Before discussing an individual's awareness of belonging to a civil society, the very concept of a society needs clarification: a society is a group of individuals. It can be a family, an extended family, a group or a village, a region of a land and beyond. Etymologically, terms for social groups denote extended families; "families of the earth"[12] is the biblical expression for it. In the modern world, societies are broken down into parts by the use of diverse criteria, so that by one criterion social units are tribes or nations and by another they are social classes, political parties or clubs. In other words, an individual may play diverse roles in society, and these may be alternatives and they may be complementary.[13] Borrowing techniques—from a neighboring tribe, party or club—is much simpler than borrowing some more abstract parts of their culture. Intellectual, religious or social leaders of a tribe, usually do that the use of items from two different systems of thought. It is possible to identify items from different cultures but with one role. It is thus possible to identify all goddesses of beauty: Inanna, Astarte, Ishtar, Aphrodite, Venus, Freyja, perhaps also Esther, Susana, Maria Magdalena and the latest movie star who serves in that role. Of course, these differ in different cultures, as the very views of feminine beauty are diverse. Somehow, this receives stress in efforts to differentiate and a gloss over in efforts to stress the siblinghood of humanity.

Nowhere is this idea more arbitrary than in discussion of the hostility between religion and science. The advocates of this hostility stress every difference between them and their adversaries that makes them look better, not only on matters that signify but also on texts that have no practical import. For example, the religious scientists cite passages from the opening of *The Book of Genesis* that agree with contemporary science and the scientists of the anti-religious persuasion do the opposite. This has a lesson for us all: when reading a text, do notice that what authors omit may signify no less than what they stress. More important: when you write a text, what you omit from it may deserve notice, together with an explanation of its omission. A writer, even a potential writer, must notice that some texts are reader friendly some not, and then find the reader-friendly author enviable and their techniques worthy of efforts to emulate them.

Reader-friendliness requires the right attitude, that of taking the reader into confidence and avoiding bamboozling them. Trusting readers to be friendly and intelligent requires resistance to prevalent training to do the opposite. To do that, it is helpful for authors to tell their prospective readers why they are taking the trouble of writing their texts. Consider a paper to be an open letter—as was intended when Robert Boyle instituted the scientific paper at the time of the scientific revolution. It is much easier to write a letter to a known prospective reader than to anonymous ones. Beginning a text with a description of its prospective readers, then, helps writing it, as does the description of the purpose of writing that text.

We tend to minimize the intellectual differences between them and us in which they are superior to us, simply by adopting their superior characteristics, and we do so tacitly. This tends to reduce the distinction between them and us. To increase the import of the distinction between them and us, we make it hard to overlook ritual differences between our traditions—even small ones. Differences in ritual stand out, especially those whose very role is to sharpen the distinction between them and us (and display our superiority to them): "the Egyptians might not eat bread with the Hebrews; for that is an abomination unto the Egyptians."[14] Bertrand Russell wrote a story, "Zahatopolk"[15] in which he describes a tribe that displays to its youths its superiority over its neighbors by showing that their neighbors eat peas. Silly as this way of thinking sounds to us, we must notice that it is a general practice. As Ernest Gellner has observed,[16] we are quick to ridicule the weird customs that they practice, but those that our own tribe practices these days look to us normal. The diversity of customs was always a part of general knowledge; the custom of approving of our customs while ridiculing their customs, however, was great discovery and a moving one at that. The ancient Greek historian Herodotus reports[17] the idea of all humanity belonging to one kingdom; he also observes[18] the surprise that some Greeks and Persians met when they discovered the symmetry between their views: each viewed with horror the burial customs of the other. Yet it was the Greeks, not the Persians, who drew far-reaching conclusions from this discovery. It is not clear what individual person made this discovery—the greatest discovery in the whole of human history and the one that as far as we know has appeared only once in human history: in Greece. It is this: since for them we are they, our sense of superiority over them is on the same footing as their sense of superiority over us; hence, it is questionable. Greek intellectuals found it not very difficult to surmise this. For, the intellectual superiority of their neighbors was too obvious to ignore: their intellectual leaders were deeply aware of it. The Greeks learned reading and writing from their Hebrew neighbors (not the Israelis, incidentally, but the Phoenicians); they learned astronomy from the Babylonians and geometry from the Egyptians, not to mention shipbuilding and metal mining and more from still others. One discovery is definitely Greek. It is philosophy, the idea that no culture can claim superiority, at least not as a matter of course. It is the idea that all humans belong to one tribe: all humans are brothers and sisters. It is the idea traditionally known as the brotherhood of man; my own term for it is, the siblinghood of humanity.

Strange! The idea of the siblinghood of humanity is ubiquitous, yet almost everywhere it is marginal and in conflict with prevailing opinions; only in Greece—or rather in Athens—it was prominent. Even in the Abrahamic religions, where it is conspicuous enough, even there it is not prominent, as it competes with the older idea of us as the chosen people. The unity of humanity is more conspicuous in Buddhism than in the Abrahamic religions, since it does not preach superiority. Yet even there the idea of the siblinghood of humanity is not dominant, so that there is no conflict between the adherence to Buddhism and the participation in any other religion. The Greek idea of the siblinghood of humanity conflicts with the Abrahamic religions head on. The conflict between the idea of the siblinghood of humanity and the idea of the chosen people is notorious: it is the conflict between Western philosophy and Western faith. Efforts to overcome it usually appear as the claim of my religion for universality. This is particularly obvious in the case of the church that calls itself Catholic, since "catholic" is synonymous with "universal," as well as the church that calls itself "Unitarian." Generally, philosophers distinguish between Universalism and Particularism, calling the one "natural religion" and the other "revealed religion"; they usually argued that the two are not in conflict. This raised the question, why is my religion better than yours is? A satisfactory answer renders one religion universal. Which is it? One answer appeared that is satisfactory: that of Gotthold Ephraim Lessing. *Nathan the Wise* says, we do not know.[19]

Taking the idea of Lessing seriously,[20] not only as a ploy to dismiss just criticism, we should also say how we treat the claim, prevalent in our religion, that we are the chosen. The answer is that we are chosen not in order to bestow on us any privilege but, on the contrary, to impose on us certain duties—usually the duty to propagate the very idea of the siblinghood of humanity. What matters here is the way Lessing makes us view the conflict while rejecting any attempt at a conflict resolution. This way the conflict has the greatest consequence possible for the philosophy of life. Which is the subject matter of this book. It is obvious: doubt invites toleration. As Peter Ustinov has put it: "Beliefs is what divides people; doubt unites them."

Interestingly, in the middle of the twentieth century a slogan was widespread in the United States of America: "go to the church or the synagogue of your choice!" (The absence of a reference to a mosque or other kinds of houses of prayer was due to their marginality there at the time.) This seems strange, since it sounds paradoxical: it advocates religion while it belittles all religious doctrine by dismissing all doctrinal differences; yet it is a concession of Universalism to Particularism—or the other way around: it is an impermissible compromise.

The Greeks have put the matter this way: there are two kinds of truth, truth by nature and truth by convention: one receives recognition everywhere and the other receives recognition only in our tribe, the one is universal and the other is particular. The Greek philosophers declared the former binding and the latter not. In other words, the wise ignore all fashion, all conventional wisdom, all that tradition sanctifies. This is (allegedly, of course)[21] the philosopher's decree: Ignore fashion! Live in the bosom of Nature! Live in the light of reason! Respect science and only science!

The above is a repetition of the decree of Greek philosophy—some say, of philosophy as such. Its wording here is different, as if the versions differ only verbally, being identical in content. They are not quite. The first variant is least problematic. Ignore fashion! This means that you may behave in unfashionable way, for example, you may stick to yesteryear's clothing fashion and dress out-of-fashion. This is good, since it frees you of some duties that you hardly need follow, while allowing you to concur with it—if and when your own taste invites you to do so. Some items matter more. Marcel Proust mentions the high-class woman who elopes with her footman to find herself ostracized when she returns to the fold. Proust does not object; he notices that she would be better off expecting the ostracism than letting it surprise her. My point is different: my wish is to represent the central idea of Greek philosophy that many philosophers adopt or violate unawares, and which makes academic philosophy and the philosophy of life identical.

The second variant mentioned above—Live in the bosom of Nature!—is problematic. By (the Greek) definition of the law of nature, we cannot break it: when we take a statement—say, Isaac Newton's theory—to represent the law of nature and we find an exception to it, we cease viewing it as a law of nature. Hence, if living in the bosom of Nature means do not break the law of nature, then, we all do that willy-nilly. Yet the image we have of life in the bosom of Nature is life in the country rather than in the city. The advice that we all live in the country is impossible to follow: there is not enough space on Earth to permit country-life for all current city-dwellers! The tremendous appeal of the countryside for most city-dwellers is just there: nothing will convince them to become country-dwellers; they wish to go there for a vacation.

The third variant—Live in the light of reason!—looks the most obvious. This is a slogan: it means, live according to science and only science. Not that we all know what science says. On the contrary, no one is an expert in all branches of science. The slogan is obvious in the sense that we can always consult a standard textbook or an expert. By contrast, consulting an astrologer or a religious leader or a political leader or a social worker will not lead to the unanimity that consulting a scientist allegedly will. Hence, the slogan is allegedly unproblematic. This allegation is positivism. It is false.

Not so fast! Science engages many people in research. It does so because it still does not answer all the questions that we ask. This is indisputable: consulting astrologers raises no trouble, as they are willing to answer every question that they comprehend. Scientists, by contrast, admit that they leave unanswered most of the questions that they encounter. Hence, we should tone down the decree, Live according to science and only science! We should do so thus: whenever possible, live according to science and only science. What then should we do when we have a question that science does not answer? Take an obvious example: an incurable disease. What do we do about it? A popular view declares[22] that a physician has to prescribe something no matter what. What a physician prescribes without the aid of science is arbitrary or traditional.

The early seventeenth century philosopher René Descartes, deemed the father of modern philosophy, was the first to take positivism sufficiently seriously to try to live by it. He was the first to try (unsuccessfully, need one say) to follow the positivist decree. He said, he had rejected all that he knew with the intent to follow only the scientific, namely, only the proven. He asked himself, what he should do before science had instructed him how to act; he decided to act in the interim conventionally. These days, positivists do not face the question openly and so they neither endorse his brave decision nor reject it.

Here is a very important corollary for the philosophy of life: those who talk tough often do so in effort to conceal their cowardice: in the words of the old English proverb, barking dogs never bite—except that at times they do, so that we should stay alert. Another important corollary for the philosophy of life is this: positivism causes a cleavage between academic philosophy and the philosophy of life. We should therefore beware of it; we can refuse to consider the versions of academic philosophy that deviate from the philosophy of life, since their aim is to expose philosophy as barren and pointless and, indeed, they may succeed to make it so. They often do so by following Descartes's demon. This demon is a part of Descartes's thought experiment. He decided, we remember, to endorse only proven ideas. Now most people trust their eyes and assume that their experiences are reliable, at least when we are careful not to allow sense-illusions to deceive us. This is naïve realism. Descartes showed that the proof that naïve realists offer is not watertight: it is conceivable that a demon deceives me and places me in scenery that looks genuine but is not.[23] Now Descartes was never so foolish as to consider this option seriously: he thought it out in order to refute the identification of naïve realism with scientific realism since he considered science identical with provability. As long as you deem science certain, Descartes's demon will continue to bedevil you; since after Albert Einstein no one any longer deems science certain, Descartes's demon has become unemployable: the academic philosophers who still struggle with Descartes's demon are fighting windmills.

NOTES

1. Rh refers to Rhesus monkeys, as its discovery involved use of them in a laboratory. Blood has many characteristics; in medicine only the practically most significant are most familiar to the public; these concern blood transfusion. Rh positive is the most common blood groups. Having Rh-negative blood group is not an illness and usually does not affect its owner's health, so people discover that they belong to that group only on rare occasions, usually in some emergency. For transfusion the most desirable kind of blood is o-Rhesus negative group.

2. Social anthropology Malinowski-style centers largely on kinship systems and their interactions with other social institutions, usually the prerogatives of relatives.

3. Views whose advocacy makes them true are known as self-fulfilling prophecies. Suspicion is the paradigm case, although it involves no prophecy and no prediction.

4. The traditional desiderata of a classification—known as essentialist definition—forbid this. It must be a partition, namely make every item of a given group belong to exactly one sub-group.

5. Martin Gilbert, *First World War*, 1994.

6. Buddhism is an exception, assuming that it counts as a religion in the first place; quite a few experts deny this, since it declares itself compatible with affiliations to other organized religions.

7. The assumption here is that backward countries rich in natural resources are poor since most of their citizens are.

8. The idea was ridiculed in the famous song "He is an Englishman" in *HMS Pinafore* of Gilbert and Sullivan.

9. In *À la recherche du temps perdu* Marcel Proust emphasizes this. He ridicules there the retention of defunct titles. His mother was Jewish. The two Jewish women in his novel had been ladies of easy virtue.

10. A few studies named *Modernization and Its Discontent* echo Freud's 1930 *Civilization and Its Discontents*, 2010.

11. Theodore Herzl narrates stories that come to illustrate his hostility to the sense of well-being, including the easy overcoming of food shortage. Aldous Huxley first opposed the sense of comfort, contrasting it with the Roman sense of virtue. His later approval of psychedelic drugs made him change his mind radically. His *After Many a Summer* and his famous *Brave New World* are emphatically ambivalent.

12. *Genesis* 12:3.

13. Michael Banton, *Roles: An Introduction to the Study of Social Relations*, 1965.

14. *Genesis* 43:32

15. Bertrand Russell, "Zahatopolk," 1954.

16. Ernest Gellner, *Cause and Meaning in the Social Sciences*, 1973, Ch. 7, "Sociology and Social Anthropology."

17. Herodotus, *Histories*, vii/19, 1977.

18. Ibid., ix/86.

19. Gotthold Ephraim Lessing, *Nathan the Wise*, 1779.

20. This is taking the autonomy of the individual for granted. As Karl Popper noted (*The Open Society and Its Enemies*, 1945, Ch. 11), some but not all religions are illiberal. (He posed there as a liberal Christian.)

21. The paradigm case is René Descartes who vowed to go on a pilgrimage to a famous shrine after he would end his experiment in total doubt. Søren Kierkegaard loathed his rationalism but admired his courage.

22. Rose Laub Coser, "Authority and Decision-Making in a Hospital: A Comparative Analysis," in Eliot Freidson and Judith Lorber, editors, *Medical Professionals and the Organization of Knowledge*, 1972.

23. See René Descartes, *Œuvres de Descartes*, 1996.

Chapter 2

Universalism versus Particularism

Eurocentrism is a new swearword that has appeared only after World War II, to win popularity at once and replace older popular swearwords: first imperialism and then white-man's burden. The individual who gave this swearword its current popularity is perhaps the currently most popular critic of Eurocentrism, Edward Said, author of the best-selling *Orientalism*.[1] This work is vulgar: it declares the whole of European literature on the Mid-East as a caricature with no redeeming features. It is obviously a sensational exaggeration. This is not obvious, as it is an attack on an (alleged) exaggeration. Its very presentation is a wholesale attack on the field as a sensational exaggeration. It is, Said declared, in part sensationalism and in part indiscriminate wall-to-wall condescension toward all non-European cultures as such. There is no need to read Said to know that almost every literature includes serious and unserious items. Hence, it is as nonspecific as possible! His success is not in the message that he has delivered but in the sense of guilt that he has provokes in quite a few of his Western readers, those who regret the dearth of their knowledge of the Middle East and its culture. Nor is there any need to know that not all general assaults on ignorance and other deficiencies can be easily justified.[2] This way of provoking our sense of guilt is a critique of the Westerners who concentrate on their own culture whilst feeling that their culture is best. This happened to intrigue me personally, since in my books Eurocentrism may count as a source of pride, although my education is not Eurocentric (it is Eastern European Jewish orthodox). Commentators who are hostile to Eurocentrism will condemn not only myself but also you, my dear reader: since you are reading this book, they would rightly observe, it is very likely that you are a white, middle-class youngster, educated mostly in European culture, generally ignorant of non-European cultures as well as of almost everything that belongs to them. Of all these qualities, one is truly likely and very significant: your European background. Only one point in the criticism carries any weight: we all are too ignorant—in general and thus also of all non-European cultures (even for those whose native culture is non-European). If this is not true of you, my reader, then most likely someone who wishes to help you in cutting your teeth on something European has advised you to throw a glance at this book. If you have read the front material of this book and the first chapter of it and you are still reading, then this abodes ill of you: the bug of Eurocentrism has already infected you, alas, perhaps incurably so. Beware of it: Eurocentrism is a particularly vicious version of narrow-mindedness!

Let me confess: the threat of Eurocentrism never engaged me, possibly since in my view of my background is Jewish culture that is only in part European: the Talmud is a major portion of Jewish culture yet most Europeans, including Jews who excel in European culture, are rather ignorant of it. Some Jewish scholars, such as David Daube,[3] are famous for their having made excellent contributions to the literature of both Talmudic law and Roman law; they are extremely rare, but then quite generally scholars of any color are. Nonetheless, it was always my intention to pay the critique of Eurocentrism its due; the idea appealed to me that in this discourse, it is wiser to exaggerate on the side of the critique of Eurocentrism than on the side of the adherence to it. Consequently, I tried to look around. My interest in music (despite of a lack of proper musical education) absorbed my effort to pay the critics of Eurocentrism their due by paying special attention to non-European music. I had little success. I found some non-European music terrific, mainly Indian ragas, perhaps also gospel music and delta blues—if and to the extent that these items count as genuinely non-European—as well as folk music of different nations, including particularly the use of diverse Asian scales and African polyrhythmic drumming. Wonderful as diverse non-European contributions to music truly are, they do not come close to the high level of art that European classical music displays ever since Palestrina and Lasso composed their immortal works before the year 1600. Many non-European composers whose works are remarkable display sufficiently strong European influences to count as European. Indeed, quite generally, European cultural dominance has given way to cultural globalization—at least in highbrow art, especially music. This gave rise to the broad criticism that is complementary to Eurocentrism: the superficiality of the global village. How does one avoid both Eurocentrism and globalization of culture? I have read composers who performed research with the intention to free themselves of both; I have found their efforts largely futile.

The picture with the plastic arts is slightly better than with music, though. The plastic art of some non-European traditions is more impressive than their music, yet none has reached the beauty and depth of Renaissance plastic art. In this respect, influence has led to experiments that had some interesting results, especially in the influence of modern paintings on artists thoroughly imbued with their own traditions. How much value this kind of experimentalism has is not for me to assess. If the plastic arts will continue to grow, then no doubt these experiments will leave their impact. How much and whether to the good is anybody's guess. The same goes for literature, although on this my view is more tentative since my impression depends on the judgments of translators; these may very well be infected with the bug of Eurocentrism. To analyze a modern novel and see in it the traces of modern Western literature and traces of native traditions requires much background knowledge that is beyond the abilities of any single observer, certainly beyond mine. Even the influence of the cinema on local culture is a vast, fascinating field, inaugurated by those who are enthusiastic Eurocentrists sympathetic to some local cultures.[4] This may very well annoy the critics who devote their time to the condemnation of Eurocentrism as a pernicious version of narrow-mindedness. They are indeed prone to some version of narrow-mindedness, since they usually tend to belittle science and technology, since these fields are the most characteristically Eurocentric: the very idea of science and science-based technology and thus the industrial revolution itself, all these great transformations belong distinctly to European culture. As it happens, not only the critics of Eurocentrism, but also some much-respected cultural leaders in the West declare science devoid of cultural value.[5] Some of these cultural leaders are adherents to Eurocentrism and some of them are strong critics of it. Laying great stress on science and science-based technology assures a bias in favor of European culture. This offers cultural leaders hostile to science an opening to criticize Eurocentrism. Their being cultural leaders steeped in European culture provides them with the opposite initiative. Western culture does not display any uniform bias: some leading Western thinkers rebel against it, some of them with no reluctance endorse it, and most of them are naturally in between.

Consider one leading Western twentieth-century philosopher, the thoroughly Nazi Martin Heidegger. He was as Eurocentric as possible. He took it for granted that his craft, philosophy, is the flower of both civilization and culture, and that only fluent German speakers can master it. His major thesis is that scientific progress has no cultural value; that if anything it is somewhat detrimental to cultural progress. He declared that he was not opposed to science and technology and that he even recognized scientific progress, but that he was seriously opposed to the idolization of science that leads to the neglect of real culture. There was a fierce debate about the weight of his adherence to his Nazi affiliation had for his philosophy in general. For, many of his admirers are impeccable anti-Nazis. Yet he was clear about it: his dislike for technology was checked by his recognition of the need of "the" party or nation or what-have-you for war machinery and thus for some hard technology. His basic philosophical thesis was the claim that there are two kinds of truth, poetic and scientific, and that the poetic is superior. Clearly, his claim for racial superiority over his Jewish co-nationals was true only poetically.

The hostility to science in Western society is not limited to philosophy proper: it is obviously part-and-parcel of Western philosophy of life, clothed as the praise for simple life in the bosom of nature. Another popular twentieth-century philosopher, Ludwig Wittgenstein, famous for his (alleged) contribution to philosophy proper, mainly the philosophy of language. His philosophy of life was archconservative: he expressed doubt about the idea that science contributes to progress—seldom but systematically and with no wavering. In the philosophy of life, he was as much a follower of the Russian novelist Lev Tolstoy as could be. Unlike Tolstoy, however, he held religion in contempt. He nevertheless admired sincere religion, particularly of barely educated people.

The most pertinent text on the matter is still of the celebrated novelist C. P. Snow—Lord Snow. It was a very famous pamphlet, "The Two Cultures and the Scientific Revolution" (1959). As he noted in his own commentary on it, "it has struck a chord." It did that for an obvious reason: it made a painful observation and he said we could do something about it. The observation is false and the cure that he offered turned out to be asinine. The observation is that our culture has split into two professional groups—the artists and the scientists—that have lost the ability to communicate with one another yet they must do so as they differ on matters political that concern us all. Yet he sided with the scientist: the artists, he said, are reactionaries—he called them natural Luddites—and the scientists are progressive—he deemed them future-oriented. The cure he offered was for the West to emulate the Soviet educational system that prevents Soviet society from splitting this way. From the present vantage point, there is no difficulty judging this proposal both uninformed and silly.

Poor as Snow's discussion was, it caused a poorer reaction. The lead critic of Snow was literary critic F. R. Leavis. He protested against the very idea of Snow that science is a culture.[6] Here he employed a verbal ambiguity: Snow used the word "culture" to denote a social set of sorts, and Leavis used it to designate "the arts and other manifestations of human intellectual achievement regarded collectively," to quote the Oxford English Dictionary. Hence, Leavis took here recourse to demagoguery.

What Snow observed that "struck a chord" as he has put it, is that in his vicinity scientists and artists do not communicate with each other—for want of a common language. Now these are elites. They comprise a small minority of the national population that Snow was addressing. He said, the significant point here is one that he repeatedly made in his many writings. No matter how conservative Britain is, he noted, almost all of its social institutions are novel. The most important novelty, he added, was the rise of universal literacy. His own grandfather was a carpenter who acquired literacy in workers' evening classes for adult education. Yet universal literacy is the outcome (not by adult education but) of compulsory education. Adult education was and still is very important, yet its import is that it popularizes art and science. Minor as its success is, it is most significant. Indeed, both Galileo Galilei and Einstein said, science without popular, amateur science is impossible. Yehudi Menuhin said the same of music. Now in the sense that every social grouping has its culture, surely the social set of scientists has one too. To repeat, it is in the sense of high culture that Leavis denies science any status; he considered himself expert on all cultures as a matter of course, and yet, as Snow observed in his famous essay, he was patently ignorant of science; *ergo*, science is no culture.

It all started with Georg Friedrich Wilhelm Hegel,[7] the most significant early-nineteenth-century philosopher: civil society offers equality, but this equality is "merely formal"; here we have the contrast between the merely formal, namely, the empty of "real" content, and the real "real": civil society versus community. Now, since community is religious, religion is to gain support even from atheists.[8] This is an attempt to construct a bridge between science and religion. It fails. Science belongs to civil society. Philosophers whose hearts are in the community find it culturally valueless. This way the doctrine of the siblinghood of humanity gives way to the doctrine of the chosen people: tacitly and quite casually. Admittedly, a society can choose to replace science with religion; but it cannot do so casually without notice of it creating profound schisms.

Still, for good or ill, so far, science permeates European culture: some leading scientists are not European by birth or by domicile, yet they are European by education—regardless of whether science is deemed a culture or not. Moreover, the distinctness of Europe is not only in its science but also in its religions and its politics: it is Christian and it is democratic. Now it is difficult to say that Christianity is democratic: for centuries, Europe was obviously Christian yet not democratic in the least. So perhaps Protestantism brought about democracy. A leading sociologist, Max Weber, wrote a leading text in social science, *The Protestant Ethic and the Spirit of Capitalism* (1905); it claims that Protestantism brought about capitalism and that capitalism brought about democracy. His claim is questionable since capitalism evolved in early Renaissance Roman Catholic Northern Italy. Gellner has argued that civil society predated both capitalism and science. His view makes sense: particular doctrines stand in the way of science and of democracy. The recognition of citizens' rights is egalitarian and it is the root of civil society. As Gellner describes it, citizenry opened the floodgate to the modern world. To cut things short, we may notice that the most superficial view has priority in subtle disagreements. Superficially, modern technology has separated the west from the rest of the world and set it apart; this then gave rise to Eurocentrism. Superficially, it was the machines, the robots, the information highways. These cannot exist without books. Karl Popper illustrated the superiority of the book over the machine by the following thought experiment: had we lost all books, we would be unable to tend the machines, and so we will eventually lose them too, whereas losing all the machines we could use the books to rebuild them and we might do just that.[9]

The beginning of explicit Eurocentrism was in the very attempt to do the opposite: it was the attempt to place Europe within the family of nations, to view it as a part of the human race. It may that has the same rights and duties as other parts do. It should have begun with Herodotus, who reported, we may remember, the great discovery of the siblinghood of humanity. Oddly, he ascribed it to the Persians: he said it began when the Persians expelled their magi from their cities. Perhaps this is not so odd, since Persia was at the time the great enemy of Greece that threatened the very independence of the Greek cities. In any case, Herodotus missed this point. Throughout the ancient Western world and throughout the Western Middle Ages, Athens, Rome, and Jerusalem competed for intellectual priority: they all drew the attention of intellectual leaders: Jerusalem due to its religion, Greece due to its science, and Rome due to its worldly, political and military strength. All were distinctly European. (Ignore the location of Jerusalem in Asia! Ignore the role of the Fertile Crescent in the very beginning of civilization!) Yet Eurocentrism was still not on the books. The beginning of explicit Eurocentrism was in the very attempt to do the opposite: a new science that began as an explicit anti-Eurocentrism altered all this: it was Bronislaw Malinowski-style social anthropology. The qualifier "social" is in contrast to the older "philosophical" that means scientific and that characterizes eighteenth-century study of the Science of Man. It included of course views of science and of social science that are no longer in fashion. The qualifier "social" is also in contrast to the "physical" namely biological. The first social anthropologist was Lewis Henry Morgan.[10] He studied contemporary American Indian society, with the most anti-Eurocentric concern possible, and yet he asked the most Eurocentric question on the books: why are indigenous Americans less developed than the immigrants from Europe? Perhaps this is how Eurocentrism came to be on the map. The answer that Morgan gave to his question, incidentally, was the supposition that all societies move on the same conveyor belt, but not on the same timetable. He used this supposition to study the social structure of ancient Greece. This is clear from the very title of his path-breaking book.

Morgan did not start a research tradition. He influenced, however, authors who did this: Franz Boas in the United States of America and Malinowski in Great Britain. They shared Morgan's problem generalized: why do societies differ? What troubled contributors to this literature was not so much the difference in kind but the difference in quality. For, it takes no effort to see the superiority of European culture over all others. This, however, is the very sin of the Eurocentrists. All modern social anthropologists have as the first item of their training the shedding of their own Eurocentrism, allegedly the chief purpose of the ritual of fieldwork that is the joining of some tribe or tribe-substitute for a sufficiently long time to learn their language and customs no matter how superficially.[11] A part of the successful outcome of this training of anthropologists is the denial that primitive people are magic-minded.[12] Anthropologists know of course that primitive people practice magic constantly. What they deny is that the expectation behind magical practice is that it is successful. The proof of this is that magical rituals obviously do not work. To see this one may take a cue from the European experience: many practicing Christians are atheists at heart. The question thus remains: why do European practice religion and primitive peoples practice magic? The assumption behind this question is that religion is superior to magic. Hence, European culture is superior to that of the rest of the world.[13] Of all the very many differences between any two traditions, this difference is what troubled anthropologists: how should they present the difference between Europeans and others without suggesting that the Europeans are superior. Anthropologists had to deny the European superiority, but only tacitly, since it is so obvious. This is no easy task.

The problem appeared first in the scholarly literature; yet it could not avoid reaching the general educated public. It solved it with a true flare-up, by a once-famous 19-page 1982 testimony to the US Senate Subcommittee on Employment and Productivity. The testimony itself drew American public attention to the Japanese working style. Consequently, for a brief timespan, Americans enjoyed the luxury of admitting some superiority to non-Europeans. The flare-up indicates the thirst of American public opinion for evidence that draws attention away from Eurocentrism. Japan displayed many ways of excellence, but that excellence is largely due to the imitation of Europe that Japan practiced vigorously at the very start of its opening up, with the Meiji Restoration (1868). It was the industriousness of the Japanese that caught the fancy of the educated public in the United States, and because it is supposedly indigenous in origin. (I do not know if this supposition is true, nor how it is possible to examine it; but this is irrelevant to this story.)

The 1982 story had a 1928 precedent, the immense popularity that Margaret Mead's best-selling study of sex education in primitive societies[14] enjoyed in the United States at the time. Her image of Samoa is idyllic. It is easy to dismiss by considering its clash with the official documents that comprise an appendix of that very book. The public ignored this. No matter how different Samoa is from Europe or the United States, Margaret Mead pointed at the superiority of Samoa in 1928. According to her story, Samoa was then devoid of the crises that European-style education incurs on teenage girls. Indeed, many a reactionary political writer declared that loneliness and the strife that accompanies it ("alienation" is the catchword for these phenomena) characterize the modern way of life and are absent from the traditional, communal way of life. This is silly. Many a writer only insinuates it by declaring loneliness and strife characteristic of the modern world. Modernity is a characteristic of Europe—more of the United States than of Europe, yet for the present discussion suffice it to note that American culture, especially its yen for the ultra-modern, is almost all of European origin. Hence, perhaps the critics of Eurocentrism express a hostile attitude toward modernity. Even so, however, modernity is not the only characteristic that the critics in question wish to reject. Is modernity the most objectionable characteristic of Europe? Obviously, the most significant target of Europe, the cause of the traditionally most serious objection to Europe, is its traditional violence: its colonialism. The latest most popular condemnations of Europe are the studies of colonialism.[15] Its popularity rests on the strong, passionate reaction to the injustice that Europeans have committed all the places that they conquered and the equally passionate denunciation of colonialism as such. In particular, Said declared the orientalist literature condescending, suffering from a Eurocentrism that he presented as so very deep-seated as to be incurable. This move discourages rather than encourages the fight against the evils of colonialism, of course. Yet it is the cause of the popularity of Said's book in the colonialist countries that he denounced. The obviously right moral attitude is to encourage the poor countries of the world to emulate the characteristics if Europe that will help pull them out of their indigenous poverty while supporting pluralism: a culture can adopt alien characteristics while enhancing its own uniqueness. It can do so in many ways. The best response to anti-Eurocentrism is pluralism; it is the wave of the future.

Once pluralism becomes a dominant part of culture—European or other—it will be possible to say a word for colonialism without fear of seeming an advocate of it. The simplest way to do this may be a word of praise for the story teller Rudyard Kipling (1865–1936), the most ardent and most vocal sponsor of imperialism. He was terrific in many ways, and his support of the cause he was advocating—imperialism—was always very interesting, even though that cause always was untenable and it still is.

Yet pluralism wants some discussion. We had better make sure that human sacrifices and burning widows will not be find comfort in pluralism even though they comprise parts of some culture or another. This, incidentally, is perhaps the difference between pluralism and post modernism. It is the difference between what Popper called closed and open societies: closed societies lock themselves in taboos and refuse to improve, whereas open societies encourage criticism and invite reforms. To take one bold step further, pluralism opposes magic and welcomes science—however exactly one may view science. Indeed, significant as science is, science-based technology is the factor conspicuous in its absence in traditional society. This has led twentieth-century political thinkers as different from each other as Dr. Sun Yat-sen and Heidegger to seek ways of benefitting from science-based technology while limiting the influence of scientific ideas on their societies as much as possible. This is impossible, as the influence of discoveries on society has its own inner logic. Surprisingly, the most practical scientific discovery, the politically most influential one, belongs not to physics and not to chemistry but to geology: it is the discovery of the dinosaurs. This discovery led to agreement beyond dispute that one day the human race can become extinct. It became obvious then that some but not all possible causes of such extinction are beyond our control. The rest is history.

NOTES

1. Edward Said, *Orientalism*, 1978.
2. Allan Bloom, *Closing of the American Mind*, 1987, is perhaps the paradigm of the recent complaint on the recent (alleged) decline of literacy.
3. Bernard Jackson, ed., *Studies in Jewish Legal History, in Honour of David Daube*, London: Jewish Chronicle Pblns, 1974; also published as a special issue of *J. Jewish Studies*, 25, 1974, 16–34.
4. Ian C. Jarvie, *Philosophy of the Film: Epistemology, Ontology, Aesthetics,* 2003.
5. A. L. Kroeber, *Civilization and Culture*, 1962.
6. F. R. Leavis, *Two Cultures? The Significance of C. P. Snow*, Introduction by Stefan Collini, 2013 (1962).
7. Karl Popper dismissed Hegel as "a clown" (*The Open Society and Its Enemies*, 1945, Ch. 12, I). See also Joseph Agassi, "Hegel's Scientific Mythopoiesis in Historical Perspective," in R. S. Cohen, R. M. Martin, and Merold Westfal, eds., *Studies in the Philosophy of J. N. Findlay*, 1985, 445–458.
8. This is manifest in *Missa Solemnis,* the religious music that atheist Ludwig van Beethoven composed as drama.
9. Popper probably never published this thought experiment. He descried it in his university lecture courses.
10. Lewis Henry Morgan, Ancient Society, or Researches in the Lines of Human Progress from Savagery through Barbarism to Civilization, 1907 (1877).
11. Ian C. Jarvie, *The Revolution in Anthropology*, 1964, is a detailed, critical discussion of Malinowski's tradition.
12. Magic practices accompany narrating myths. The narratives are easier to dismiss than practices.
13. The wish to deprive Europe of this superiority led to the view of it as a matter of civilization in contradistinction to culture; this is why the distinction that Ferdinand

Tönnies has made, between community and society (*Gemeinschaft und Gesellschaft*, 2019 [1887]).

14. Margaret Mead, *Coming of Age in Samoa: A Psychological Study of Primitive Youth for Western Civilization*, 1932 (1978); Paul Shankman, *The Trashing of Margaret Mead: Anatomy of an Anthropological Controversy*, 2009.

15. Franz Fanon, *The Wretched of the Earth*, 2007 (1961); Edward Said, *Orientalism*, 1979.

Chapter 3

Background to Radicalism

The concern that radically inclined individuals display is noble. Indeed, the very first aim that they usually undertake is to combat narrow-mindedness. The second is to present an ideal agenda. This is noble, but unwise. It is, to say the least, questionable. It is a very difficult task, as the narrow-minded have no trouble ignoring any attack on their narrow-mindedness or dismissing that attack on some pretext. Moreover, it is very easy, and also common, to hide narrow-mindedness behind an unattainable ideal. On the face of it, the better attitude toward the narrow-minded seems to be to leave them alone. Indeed, this is what most people do. Radicals find this attitude unserious, since, obviously, the narrow-minded are dangerous—politically and socially, even when the ill they do is mere deficiency, the outcome of the unnoticed absence of any opportunity to broaden one's horizon. This deficiency is famous: its name is provincialism. The radical claim that to ignore narrow-mindedness is unserious is itself unserious: those who have nothing better to do can always seek some wrong cause to attack. Moreover, what harmful deficiencies invite is seldom criticism: usually it is education. Discussions of the way to improve education in order to reduce ignorance and narrowmindedness or any other prevalent ill is most welcome. Admittedly, all too often proposals to reform education meet with opposition from the uneducated and narrow-minded. This attitude does invite criticism, yet to get the critic heard is very difficult. Poor countries need education in order to pull themselves out of poverty.[1] Their leaders often recognize this only partly: they welcome vocational training. Vocational training is not sufficient to remedy ignorance. To pull themselves out of poverty they need education proper, especially the ability to criticize traditions and improve on them. The most conspicuous impediments surround the discrimination toward women. Poor countries remain poor mainly because their leaderships prevent all moves that may lead to women's equality, particularly the education of girls. This absence of equality is not mere provincialism. It is pernicious narrowmindedness and it is quite different from its run-o-the-mill manifestations; it is dogmatism or even bigotry. It invites us to seek means to overcome resistance to education. The usual expression of this resistance is as the view of one's culture as superior. This superiority may be real, and then it is particularly difficult to overcome the resistance to education. This does not make sense; being well educated (in one area) is no reason for resisting more education (in other areas). This was the advice, for example, of misogynist Friedrich Nietzsche: he advised educated women to oppose women's education—allegedly in order to maintain their advantage over most women.[2] Wherever discrimination reigns, not only the incumbents support its continuation; the better off among the discriminated demanded to keep their own superiority such as it was. This way they the system maintains a peculiar version of obscurantism: narrowmindedness. Only legislation and education might help here—usually their combination, namely legislation that reforms the education system to make it a tool against obscurantism and bigotry.

To argue against obscurantism and bigotry still is a waste of time, since the afflicted would not listen come what may; even when they try to listen, they usually hear the valuable messages in distorted fashions that render them valueless. It is still a very important task to fight obscurantism by other, more effective means. These still are mainly two: education and democratic legislation. In both of these cases, however, the most obvious task is to prevent the obscurantists from spreading the ideas that support their narrowmindedness, while spreading the ability to reveal broadmindedness—both as general knowledge and as an ideology—the ideas of intellectual and moral autonomy—naturally with the aid of the national education system, no less. Now unfortunately, even the best national educational systems around are not sufficiently devoted to helping the young to develop their autonomy. The paradigm case is the battle against creationist biology: it focuses on a battle for Darwinist dogma. The most serious and difficult to combat version of narrowmindedness is scientism—the dogmatic clinging to the last word in science. Thus, the dreadful debate over the teaching of Darwinism is the battle between teachers of dogmatism: should we teach Darwinism or creationism (whatever these labels stand for). The obvious answer is, both. Yet this answer has no advocate in the educational systems in which this debate has raged.[3]

This is so for an inherent difficulty: teachers are supposed to instruct, and it is easier to instruct by indoctrination than by appeal to autonomy, since the autonomous student may oppose the teacher. Thus, as long we consider the task of schools as serving as means for instruction, they are not and cannot be the means for spreading enlightenment that they should be. This is no news. The father of the modern enlightenment movement, the Elizabethan thinker Sir Francis Bacon, hated schools and declared them fonts of prejudice and superstition. The leading ideologist of the Enlightenment Movement was

Jean-Jacques Rousseau. His ideal enlightened man, the hero of his *Emile*, is self-taught: even literacy he teaches himself, which is most unlikely.[4] It is no good to try and rewrite this novel: we do not need an image of the enlightened individual; we need the minimum conditions for the avoidance of extreme obscurantism and the best means for implementing them. Even less: we need to know what the worst obstacles on the way to the autonomy of the individual are. One of these is perhaps the erroneous satisfaction with the present situation. One of the major manifestations of this is perhaps the widespread myth of Eurocentrism.

Erroneous as Eurocentrism surely is, the error in question is not Eurocentrism itself but its uncritical endorsement—including, of course, its uncritical endorsement of natural science. That it is an error should be obvious. For, openness to criticism and the encouragement of criticism is what characterizes Europe's culture most, as it is what rendered the development of natural science possible in the first place. The ability to invite criticism depends on autonomy. The philosopher Popper said, criticism is the most significant characteristic of both science and democracy, the two main aspects of all European traditions. Not all citizens of democratic countries are autonomous; not even all those who spend their lives in scientific research or in the running and the spreading of democracy. Yet what characterizes democracy—what characterizes science even more—is the prevalence of criticism there—not enough but more than in any other known human culture, past or present.

Now this still seems too good to be true. We know that the Inquisition is also characteristic of Europe; it is more European than imperialism, since empires appeared on other traditions too; it is more European than religious oppression, since religious oppression too is ubiquitous, but not inquisitions.

Yet the difference between the inquisition and other forms of religious oppression is both secondary and due to natural science. This may be questioned since the critical attitude is nowhere in full control. The opposite of the critical attitude is defensiveness. It prevents the recognition of the fact that the inquisition is peculiar to European culture and that its roots are in the scientific culture that is unquestionably European.

The same goes for the Holocaust, that worst disaster that ever befell humanity, and even more so. What is most obvious about it is genocide, yet genocide is the commonest consequence of war. The genocide in the Holocaust was science based: a leading biologist, Konrad Lorenz, praised the Nazi racist conduct in his scientific publications. The nation that devised the Holocaust and engaged in executing it was European. Germany was one of the best-educated nations in Europe. Is this then not a characteristic of Europe too? I do not know. This, however, is clear. The Holocaust was a disaster for Germany; not science. More generally, the modern world can reach exciting heights of success but also abysmal depths of failure. What they share is science and science is all good even if it is open to misuse in the Holocaust.

It is the characteristic of science that it can cause both the recognition of the siblinghood of humanity and the advocacy of world peace. Yet these are not specific to science. Eurocentrism can be both the advocacy and the condemnation of cosmopolitanism and of science.

Hence, to say that openness to criticism and the encouragement of criticism is what characterizes European culture most is not to say that most Europeans will advocate criticism, nor even that they would welcome it. As it happens, most Europeans I have happened to meet disagree with this characterization. They often consider the devotion to criticism perverse. They often say, to appreciate criticism is against human nature; no matter how just a criticism is and how beneficial it is to accept it, they say, human nature makes people reject it—at least as the default option. I do not understand this, but I report it faithfully: dislike of criticism is also a deep-seated characteristic of Europeanism—in contradistinction to most other cultures that are simply ignorant of the very debate about criticism. Indeed, discussing the place of criticism in culture is a necessary condition for critical mindedness; but it is not a sufficient condition. Thus, for centuries European learning was in the wake of Aristotle, who claimed that his views were certain and thus beyond criticism. He was in error, of course. The mediaeval world, including the European tradition then, rejected first his view that everything farther from us than the moon is immutable. When a new star appeared in the sky, astronomers ignored it, which made possible clinging to Aristotle's views uncritically. Yet China, where the new star was recorded, did not develop a critical attitude either. It did not alter Chinese astronomy.

Still, allow me to repeat, the lack of awareness of openness to criticism is hard for me to comprehend. Europeans will gladly agree that the European political regimes are characteristic of Europe. Now Europeans will gladly admit that each European country has its own regime, and that no two are identical. Yet, they will observe, quite rightly, let me add, that they all share one significant characteristic: democracy. What is democracy? I do not know; there is no agreement about it. Indeed, since the label "democracy" is laudatory, every country calls itself a democracy, including countries that most Europeans consider not democratic in the least. The situation is a bit less problematic if instead of speaking of democracy we speak of parliamentary democracy. Not that every country that calls itself a parliamentary democracy and has a parliament is a parliamentary democracy; at least there are such countries that other countries refuse to consider genuine parliamentary democracies. There is also fairly general agreement of what kind of parliament is non-democratic: a parliament with a ruler who can limit its agenda. Since a country that calls itself a democracy and has a parliament need not be a genuine democracy, we still do not know what country that calls itself a parliamentary democracy and has a parliament deserves recognition as a genuine article, as a genuine democracy. What criterion is the right means for deciding this? We still do not know.

Nevertheless, we do know the difference between active parliaments and those that function as rubberstamps only. Parliaments in which oppositions have no power to make a difference—mock oppositions or puppet oppositions—are not the genuine articles. Now, clearly, just as there are mock oppositions so there are mock changes or cosmetic changes. The difference between a genuine change and a cosmetic change, however, is not hard to draw: the genuine change comes to correct some serious defect. Finding a serious defect is criticism. Hence, the democracy that is admittedly characteristic of Europe is one that fosters criticism. This is the idea of the philosopher Popper. Many European intellectuals to whom I have put this argument have rejected it. They could not find fault in it, yet they declared it faulty; they declared it a sophism.

Sophism is an intentionally concealed logical error, used to derail an argument in the direction that the sophist approves of. Whether sophists are right or not matters little, what matters is their immorality. What makes them immoral is that they manipulate the people who wish to argue with them in good faith, thereby violating their autonomy. This leads to a famous traditional debate: is deception for a good cause good or evil? The Jesuits openly advocated small deceptions for significant good achievements—for saving the souls of the subjects of their deceptions. Their name for this attitude is a famous slogan: "the ends justify the means." Now this slogan is a tautology; the Jesuits used it to advocate an attitude that makes light of individual autonomy. Any attitude that belittles individual autonomy is paternalism: I know better than you do what is good for you, and so I should use all means I have to make you behave the right way, including manipulating you. Almost all traditions share this attitude. The exceptions are some Western traditions, chiefly the early nineteenth-century expressions of criticism of the philosophy of the Enlightenment Movement, the Romantic philosophers, who recognized the autonomy of some individuals. They called them heroes. Only heroes are autonomous and they should function as leaders. The major role of leaders is that they have the right to impose small changes in the national traditions. The problems that the romantic philosophers had faced, then, are two. First, what social group is sufficiently solid to count as a (national) tradition? Second, who qualifies as a leader? They answered these questions by *Realpolitik*: the leader proves his leadership by usurping power and utilizing it effectively. This is the cynical view that might is right. It says, contracts are worthless since they are effective as long as the strong supports them, so that it is the interest of the strong that matters, not the contract. That this idea is false idea shown by how we consider trust to be valuable and make effort to deserve it. All this is what comes with paternalism; those who support it are not clear about it. Alternatively, they are and so they recommend paternalism only for the short run, as a tool to overcome the worst and most pressing national problems. This is not the worst of paternalism. The worst aspect of paternalism is that though it comes as a mere stopgap, it perpetuates itself. It does this because it encourages submissiveness and ignorance in the people at large.

Arguing against paternalism is thus major item on the agenda of all modernizers everywhere. Critics of Europeanism and of modernism often support paternalism, although they tend to do so tacitly, knowing that most European audiences will not easily succumb to it. This is not to say that they are avid defenders of the autonomy of the individual in all cases. It seems that quite generally people are not sufficiently aware of what they stand for, of the implications of what they do, or of the leading characteristics of their activities. This raises two questions: how does that come about, and what are we to do about it? Before that, we may try to learn from the fact that European culture has acquired the ability to raise at least some autonomous people quite regularly.

The story begins with the discovery that the culture to which I belong is not privileged. Socrates claimed that he was an advocate of the political system of Athens, namely, Athenian democracy. He said that because Athens allowed for discussions of the question that in his view is the most important thing in life, namely, what lifestyle one should adopt. (In other words, what religion should one join? What profession should I choose? What interests should one cultivate? And so on.) Plato translated this question to what opinion should one adopt? What idea should one believe? His answer won almost unanimous endorsement before the twentieth century: one should believe all and only proven ideas. The question then arises, what is proof?

Before that, a criticism of Plato's proposal appeared almost at once: there is no proof. This idea goes by the name "Pyrrhonism" after Pyrrho, the founder of *the* skeptical school. The word "skepticism" denotes search or looking for, related to the root "spec" of the two words, "aspect" and "spectacle." What confuses the issue most is the overlooked fact that the sceptics endorsed Plato's recommendation: one should believe only proven ideas. Further, since there is no proven idea, one should believe nothing. For centuries, the discussion turned around the question, can one believe nothing? The sceptics admitted that it is not easy but promised a particular peace of mind to those who achieve it, one that they called *ataraxia* (from *tarachê*, meaning "disturbance: sceptics promise their recruits freedom from all sense of disturbance).

Most Western philosophers consider the task of refuting skepticism the most important task in philosophy. (The meaning of "philosophy" is Western philosophy; here is a major source of Eurocentrism in circles that tend to be philosophically oriented!) For, believing nothing, nihilism (from the Latin *nihil* that means nothing) deprives one of the abilities to fight for the forces of good against the forces of evil. Worse, skepticism gives people permission to give up philosophy, to drift into irrationalism.

Plato already declared unserious the preference for believing nothing. He was convinced that proof is possible since mathematics rests on proofs (*Meno*). This is why Euclid's *Elements* (that appeared around 300 BC) was for centuries the most popular best seller in the west, second to none but the Bible. The idea of proof that it displayed is the deduction of significant theorems—Pythagoras's theorem in particular—from self-evident assertions, from ones that a reasonable person cannot doubt. These are the axioms. This is why the very appearance of alternatives to Euclid's axioms made rationalism suffer a great setback. The axioms became objects of faith, and so the difference between science and religion broke down, and this led to the rejection of the idea that science is unique; which in turn led to dismissing the claim for uniqueness of science as merely Eurocentric.

A compromise soon appeared: take your axioms on faith and believe only what follows from them. This idea is *fideism* (from "fide," faith). It is captured in a once well-known Latin tag: "*Cum negante principia nequit disputari*" or, in vulgar American, "there's no arguing with a Red." Fideism takes it to be impossible to reason about principles. It follows then that the most important things in life are beyond reason: one makes decisions about them are without the control of rational deliberation. This pushes critical debates to the margins. It is irresponsible.

Let us stay with the problem: the views that I grew up with have no privilege: I have endorsed (namely, believed) them not because I have found them the best available; I have endorsed them merely because I happen to have been born into my tribe and not into yours. Such an attitude is not good enough: I should now, as a grown up, reconsider, and examine the question, which culture is better, yours or mine. Indeed, some parts of the small Christian sect of Baptists, do not accept children to their ranks, not even the children of their own congregation, but wait until they become adults and join the church out of deliberation. Something like that happens to people who wish to join the Church of Rome: The Catholic Church is supposed to reject them until they convince the priests whom they ask for conversion, that their conversion is rational. Somehow, this does not convince others. Whatever rational arguments are, for them to be relevant to the siblinghood of humanity they should convince all who can think rationally. What this is we do not know, but Baptism or Roman Catholicism it is not.

Plato's solution to the problem of rationality is to endorse only proven views. If some rational people claim that it is not the right solution, then by its own lights it is not the right solution. For the same reason post-modernism annoys rationalists: it threatens the very idea of the siblinghood of humanity. Except that it too suffers from the same defect: by its own lights, it cannot be superior to Plato's rationalism! Now when the two extant ideas demolish each other we are in deep trouble.

Hence, said Popper, we should go back to before Plato, to the so-called pre-Socratic philosophers: they argue that already to subject your views to criticism suffices for you to count as rational. Hence, you need not discard your culture wholesale; you can do so piecemeal: whatever in your own culture seems to you inadequate or inferior to its equivalent in a neighboring culture, you may try to improve.

This is a new attitude to tradition. The most common attitude to tradition within cultures is conservatism: preserve culture; if you must improve it, do so in small steps and slowly—in the intent to avoid disaster. The most common Western attitude to culture is radicalism: do away with it, and totally so. The attitude ascribed to the Pre-Socratics is gradualism: change tradition stepwise, by criticism and efforts to accommodate it by corrections.

Strange. Parliamentarism is the characteristic of Western political traditions, and radicalism is characteristic of Western intellectual traditions. The one is gradualist, the other swinging. People are not aware of what they are doing and not even of what they aspire to do!

How does this come about?

People know surprisingly little about themselves. This is a general discovery. That they do not know their own wishes was discovered in psychology—chiefly due to Sigmund Freud; that they do not know their own customs is the discovery of social anthropology—chiefly of Malinowski; that they do not know the rules of their own languages is the discovery of linguists—chiefly due to Ferdinand de Saussure and Benjamin Lee Whorf; that they do not know the rules of research was discovered by methodologists—chiefly William Whewell and Pierre Duhem.[5] The most important item here is, we cannot avoid all error, since we do not know that what we take for granted is error-free. In particular, since scientific proof allegedly rests on observation, it is not conclusive. Leading intellectual and moral thinkers considered this conclusion a disaster.

The scientific revolution in the seventeenth century followed rules that the Royal Society of London had laid down. That Society declared that it followed no leaders and required no faith from its members. It nevertheless followed two heroic figures: Nicolaus Copernicus and Bacon. Unfortunately, Bacon called Copernicus a charlatan. The reason was simple: Bacon recommended trusting our eyes, and he observed the fact that the sun rises and sets, an observation that Copernicus had explained away. John Locke came to the rescue: we see sense data, and motion is not among them; it is a hasty conclusion.

On this, Locke was in error: as hunters know, we do observe motion. This is not the worst of it. Whatever data are, what makes them infallible? The idea that observational data are (or can be) infallible is essential for the popular view of knowledge that goes by the name of inductivism. This philosophy rests on the most famous and most refuted psychological theory: associationism. A century of research that followed ideas of Immanuel Kant led Oswald Külpe to reject this theory. Kant himself did not deny that sense data are certain; he denied them scientific status prior to their verbal articulation. Although he did not mean to question this articulation, he led people to do exactly that. Nevertheless, most Western thinkers take for granted Locke's associationism in one variation or another. This is very Eurocentric and not very intelligent.

NOTES

1. Swiss Johann Heinrich Pestalozzi (1746–1827) conceived universal education and vocational training as means to heave Switzerland out of its poverty. He thus invented the idea of human capital that economists offered in 1960. See Johann Heinrich Pestalozzi, *Enquiries into the Course of Nature in the Development of the Human Race*, 1797; *Leonard and Gertrude*, 1891.

2. Nietzsche was always vague enough to allow diverse reading of his texts. (His aim was to impress, not to instruct.) He did say education should lead men to manliness, namely violence. and women to womanliness, namely pregnancy. Yet his texts were always open to reading it in approval or in disapproval of current affairs. See Friedrich Nietzsche, *Sämtliche Werke*, 1980.

3. This is the wave of the future: the Academic Freedom Movement in the USA advocates teaching both sides.

4. Jean-Jacques Rousseau, *Emile, or On Education,* 1979 (1762).

5. All these discoveries belong to the turn of the twentieth century, at the collapse of the modern, stable, regimes that followed the Napoleonic era in the whole of Europe, with the possible exception of pre-Haussmann France.

Chapter 4

Gradualism and Pluralism

European tradition is critical—it challenges its members to criticize their own heritage as best they can, and then try to improve it. Yet, paradoxically, its traditional, semi-official ideology is Platonic: European ideology deems the rational as the proven. Hence, it views the rational as immune to all criticism. Hence, it takes criticism and what it demolishes as possibly valuable yet as definitely a mere appetizer: criticism only shows that what yesterday we deemed rational is not. So, as the main dish, European ideology considers the genuinely rational, namely, the validly proven. It is quite unclear just what has been proven, but at least we know that the proven has survived the most serious and severe criticism that has ever launched against it. What else we require of the proven? We do not know. Now, this need not be disastrous: it is possible, of course, to have proof without proof procedure, the way we can discriminate a piece of good art from other artworks without a theory of it, without an aesthetic. Indeed, it is a historical fact that claims for proof are older than any idea of proof procedure: Parmenides viewed his view as proven—as "logon piston"—yet he probably could not say what procedure leads to proof. We do not even know how to be proven or provable signifies. The literature ascribes the idea that only tautologies are provable to Wittgenstein. This is true in detail; broadly speaking, it was Kant who found two kinds of provable assertions: the tautological or uninformative and the transcendental. He was very strict: he recognized as tautological only statements of the form "a is a" ("a ball is a ball") and "ba is a" ("a round ball is a ball"). Yet to the Ancient Greek authority Aristotle has ascribed two more tautologies—"a or not a" and "not (a and not-a)," which seemed to Kant problematic. This led researchers, beginning with George Boole and Gottlob Frege, to study tautologies in general—research that a little over one century after Kant's death led to the invention of the computer. Now, Kant distinguished between analytic logic—that we call logic—and synthetic logic that he called transcendental logic. His distinction rested on his idea of transcendental proof. It was less than fortunate. Whatever his idea of transcendental proof amounts to, the examples of transcendentally proven ideas that he gave were the axioms of Euclidean geometry and the principles of Newtonian mechanics. Since Einstein has superseded both of these, today we leave Kant's transcendentalism to philosophers who prefer to ignore up-to-date natural science. Kant had offered a transcendental proof of Euclid and of Newton: being scientific there is proof for their validity. Since empirical proof is impossible, he added (following David Hume), the proof of the axioms of natural science must be prior to experience. They comprise *a priori* knowledge. Of course, Einstein superseded these proofs; he dared do so since he relied on Hume. Under his influence, it is now standard to deny that natural science rests on proof, since proof render impossible the improvement of science through modification.

Today, then, the knowledgeable do agree: proof procedures exist only within logic, and logic is utterly uninformative about God's world. It is an outlook that scholarship has earned the hard way: modern logic grew out of the search for a clear idea of scientific proof. The search ended with a sad conclusion: there is no such thing. This conclusion of contemporary logic is not universally well taken. Especially philosophers and others not well versed in logic tend to dismiss the conclusion. Some of them say, Plato has offered what he (rightly, they add) considered a proof procedure. It is familiar as a mystic union[1] or ecstatic revelation. In the whole history of philosophy, only one idea of [non-logical] proof has ever been offered—by Plato and Aristotle, al-Farabi and Moses Maimonides, and even Bacon, the last rationalist who advocated it.[2] Anyone who ever spoke seriously of proof turned out to have no idea about it or the idea of the mystic experience or of logic.[3]

Now this seems to settle nicely all practical matters. Alas, it does not: when one professor criticizes another, the other takes offence: you accuse me of having cheated when I offer my views as having passed the stage of critical examination! In other words, the idea that under some specifiable situation anyone may speak while avoiding all error must render personal all criticism of anyone who speaks in such a situation. This is why professors take criticism personally. It is only some criticism that they resent. Before a public lecture, a professor may invite the best critics around for private sessions with the request that they offer all the criticism they have. Afterwards, in the question-and-answer period, if anyone in the audience, particularly a brilliant student, offers some deadly criticism, the speaker takes it personally, especially since the audience may laugh savagely. They do so out of sheer embarrassment, for sure, yet it hurts all the same. It is a humiliation.

To my regret, I agree with the audience: an error of a lecturer may indeed be embarrassing. When? The audience does not have a general answer to this question, but their answers in detail are generally very good. Some errors are embarrassing—because they are avoidable. Others are no less than admirable. Einstein's theory of gravity exposes Newton's theory of gravity as an error.

Some physicists to whom I said this when I studied physics in the university asked me why I hate Newton. The reason for this is the confusion of avoidable error with a commendable one. The theory of Einstein is a modification of the theory of Newton. Hence, with no Newton there is no Einstein. This is why in theoretical physics courses the theory of Einstein follows the theory of Newton. Viewing the human intellectual heritage as a progressive series of errors is not only enlightening; it is also relieving of the impossible task of doing research without committing any error. Einstein stressed this point repeatedly. It takes away the unpleasant tension that accompanies research that should be a most pleasant activity.

This tension does not stay within science. Hector Berlioz and Felix Mendelssohn were leading early nineteenth-century Western composers. They were good friends. Mendelssohn found appalling the music of Berlioz, or rather his orchestration. Both feared that this fact might pop up in a casual conversation and impair their friendship. For all we know, it did not; nevertheless, it is not hard to see that it was a pity all the same.

Does this confusion of avoidable error and commendable error invite a change? If so, is it open to change? To which change is it open? Is that change desirable? Each of the answers to these questions depends on the question, what is the proof procedure that we apply to any given answer? The Royal Society did not answer this question. It had endorsed early in its existence a rule for the selection of repeated empirical observations to print in its *Philosophical Transactions.* As to theories, the Royal Society allowed their publication as brief appendages to papers on observations. This covered most of the difficult questions as to what is publishable in the learned press. There remained one difficult question: what is an eyewitness testimony? For this, the rule is: What counts as an eyewitness testimony in law-courts counts as an eyewitness testimony in science too. Yet not all eyewitness observation is acceptable in science. To be scientifically acceptable, however, it has to be a repeated report declared repeatable. (This is why science ignores observation reports on extra-terrestrial objects.)

Repeatability was hardly uncontroversial: at the time, England was host to a literal witch-hunt by Matthew Hopkins, the Witchfinder General. You would not learn this from reading the scientific literature of the day; it took over half-a-century before English law abolished the witch-hunt. You would not learn this from reading the current literature on the history of science. Perhaps this is because most of the histories of the "hard" natural sciences stick to scientific material, even at the cost of willful distortions. It is hardly avoidable: most historians of natural science take for granted that natural science is perfect and that scientists are fallible. Hence, their errors are not science and so there is no need to include them. Except that often the way a scientific theory usually appears in the up-to-date scientific textbook is a correction of a previous scientific theory. Would it not make more sense to include both the erroneous idea and its correction as parts of the history of science?[4]

We may call such corrections revolutions. We may, alternatively, call scientific revolutions only major corrections. In addition, we may follow Thomas Kuhn[5] who deemed only major changes revolutions and then haggled about the minimum size of a major change. The history of science cannot arbitrate such disagreements.

The history of science includes stories of the discovery of many empirical observations that comprise refutations of other, earlier, empirical observations. Often, reports of observations appear as corrections of earlier ones. This is not to say that the reporters of the earlier observations were irresponsible. Their own individual observations are not in question. What is in question are their reports of these observations, since these are not the statements of individual facts but their generalized versions. These often turn out to be false. Nevertheless, they are very significant, as they are the best we have: of necessity, we observe the world; when we enter the world of science, we find that many of our observations that we had taken for granted are false. This process is endless. Still, let me admit, although what I say here seems to me obvious, scientific researchers often resent such observations. (Possibly, this resentment is due to the dogmatic rejection of some scientific innovations.) The great seventeenth-century thinker Robert Boyle laid down the rules of scientific procedures early in the days of the scientific revolution, during the rise of the Royal Society. At its very foundation, its Fellows decided to adopt his rules. These rules were operative then, but they were insufficient. Half a century later, Newton laid down further rules, especially the rule about refuted empirical observations: their refutation prescribes not their full rejections but their rewriting in *ad hoc* modifications aimed at rescuing them from their rejection. Galileo and Bacon had both proscribed the application of this procedure to theories. This still holds for theories. Newton prescribed its application to empirical observations! These rules are still operative. Their use is a testimony that science ignores proofs—contrary to Newton's own view that science rests on induction, a procedure that is not good enough but that is the best available. It always amazes me that philosophers who wish to defend science and its procedures ignore what this paragraph describes.

At the root of all Western philosophy and of natural science is the Greek idea that there are two kinds of truth: truth by nature and truth by convention: the former is immutable the latter is not. This idea survived in Western philosophy until the year 1905. Two events then occurred. First, Einstein effected a scientific revolution; earlier, tradition deemed the idea of a second scientific revolution out of the question. This raises the meta-question, what is a scientific revolution? Indeed, when a century before Einstein, Joseph

Priestley fought alone against Lavoisier's anti-phlogiston theory, his strongest argument was, it was a revolution, since at its time, the doctrine of the phlogiston was hailed as a great scientific achievement. No one took the trouble to answer Priestley. When public opinion recognized that Newton's theory of light was refuted, in the year 1818, it shook the scientific world.[6] People asked, is it possible that the same will happen to Newton's theory of gravity? William Whewell came to the rescue. He explained the difference in scientific status between Newton's two theories and thus unintentionally and inadvertently answered our question: if ever anyone will manage to replace Newton's theory, that person will affect a scientific revolution. Henri Poincaré rejected Whewell's reasoning, but stuck to his conclusion: even if there will be a ground for a revolution, he said, we would prevent it. He had an occasion to do just that: when in 1911 he met Einstein and commented on his 1905 theory (in the famous Solvay conference). He graciously gave in.

Students of scientific method agree: their (self-appointed) task is to present rules for research that avoid the extremes of dogmatism and skepticism. It is possible to view this as prescriptions: do not be a dogmatist, else you block your own curiosity, and do not be a skeptic or else you get lost. This seems to me excessive since it is better not to tell people how to behave. It is possible, alternatively, to view this as refutations: a theory of research that does not preclude dogmatism is mistaken, since dogmatism is bad for research. This alternative is better, but it is problematic too. Some researchers were stubborn and refused to learn from failure. Most of them died frustrated, of course, but avoiding dogma is no guarantee against frustration. Moreover, some blockheaded dogmatists contributed to natural science or to technology. Some dogmas that were silly superstitions helped some researchers to advance natural science and technology. (Duhem made this point emphatically: he said there is no induction, but he recognized the value of the theory that natural science is inductive—as encouragement for research. He compared it to the Northwest Passage that has provided incentive for geographic discoveries even though it does not exist.) A much more significant case is the quest for the philosopher's stone that guided all chemistry and mineralogy until the scientific revolution. Most historians of science do their best to conceal such cases, yet I must express my appreciation of their candid praise for my severe attack on them. Nevertheless, the fact remains that Cambridge University, where Newton was a professor, accepted with gratitude his scientific manuscripts but declined his religious and metaphysical ones. This is how those papers ended up in the Hebrew University of Jerusalem.

The twentieth-century philosopher Bertrand Russell, one of the greatest thinkers of all time, confessed that he failed to develop a theory that managed to avoid the Scylla and Charybdis of dogmatism and (utter) skepticism. It seems to me he was right to see the extremes as opposed to common sense but too demanding when he looked for a theory that precludes the both of them. We can always fall victim to dogma and so we should always be on guard, said his admirer, Popper. This is unavoidable for any gradualist theory. Gradualism has great attraction, but it is not popular, as it does not help make life easy. Popular theories divide the world into black-and-white. Their advocates offer simple rules by which to distinguish between them. Science is proof, follow science whenever possible and you cannot go wrong. As against this, commonsense says, whatever is too good to be true is not true. Commonsense is tentative, and what we hold tentatively we hope we hold not dogmatically and with readiness to improve on occasion.

A most conspicuous case is that of Thomas S. Kuhn. His 1962 book, *The Structure of Scientific Revolutions,* was a best-seller. It comprises less than 200 pages; stripped of its historical examples it is some twenty to thirty pages long; and it had an explosive impact. Kuhn noted that received scientific opinion changes. He took for granted that science allows for repeated revolutions. That he declared in the title of his book that promises to discuss their structure—meaning the structure that they all share. This promise he never tried to fulfill. He denied that scientific theories are refutable. Taken literally, this hardly bears examination. What his argument invokes is that it is possible to accommodate or dodge any refutation. To the logically minded, this is no news. Suppose an archeologist said, every clay utensil is in principle mendable. It is not true, but it is an interesting challenge. Suppose one words this assertion thus: there is no broken clay utensil. This sounds silly. No one speaks this way. Yet many philosophers of science emboldened by reading Kuhn declare that no [scientific] theory is refutable. Why? Perhaps because it sounds new and challenging, whereas in its proper wording—every refuted theory is capable of being rescued from refutation by *ad hoc* tinkering—it is a famous fact all students of the law are familiar with since antiquity. These philosophers often marshal the claim that the impossibility of refutation is the great discovery of Duhem and/or W. V. O. Quine. Both denied this attribution as a matter of course. The Duhem-Quine thesis is that it is always possible to reconcile any two statements, which is ancient knowledge, plus a new observation: the test of a theory takes place not of it in isolation but with the aid of auxiliary hypotheses (such as the ones concerning instruments); logic allows that the clash is due to these auxiliary hypotheses. This argument saves the theory from refutation even with no *ad hoc* tinkering. The point that Duhem made, and that Whewell had made before him, is that when a theory gains protection against refutation, it thereby loses its capacity for empirical support. When Popper repeated this in 1935, his peers (the "Vienna Circle") did recognize it as an innovation, and indeed one that they rejected. The true innovation that Popper proposed is that researchers agree on self-governing rules. The most important of these is, avoid moves that reduce the vulnerability of hypotheses to refutation. The simplest way to block this vulnerability, Popper observed, is to ignore refuting information. This sounds too silly for anyone to practice, until a president of the United States did so regularly; he even proposed this as a policy. As to science, it follows the proposal that Popper has suggested: although it forbids *ad hoc* amendments of hypotheses as a rule, it allows exceptions when these raise (rather than reduce) the vulnerability of theories. Thus, when a researcher rescues a hypothesis from criticism by redirecting the blame toward some instrument, the move may win acclaim when that researcher takes the trouble to examine the instrument and point at the source of the trouble—perhaps even offers a way to fix it.

Thomas Kuhn did not go that far. It may look as if he did since his terminology is vague. He said his quarrel with Popper is merely terminological![7] Yet he never allowed ascribing to him a clear view of anything: his terminology seems vague on purpose. What is clear in his theory is that despite his rejection of the idea that science is proof, he stuck to the idea that science is unanimity. The unanimity is not across the board, of course. The idea of scientific thinking as black-and-white remains in his philosophy, even though for no reason at all. Public opinion was wavering between the desire for proof and the feeling that proof puts the whole world in the wrong with the exception of the chosen people, those who abide by proof and nothing else.

The idea that we should abide by proof and nothing else is hard to follow since there is no proof for most answers to those questions for which we have some answers, not to mention the questions on which we have no answer at all. Consider, for example, the plain question, why is the sky blue? In physics, an answer to it first appeared in the year 1900. In that year that answer met with a rejection, which rejection heralded the quantum revolution. Five years later came the revolution that Einstein initiated, that presented the older, established theories in physics, mere approximations and special cases. Whereas Newton had endorsed the view of science as proof, Einstein view science as the gradual approximation to the truth. This idea met with a tremendous ambivalence within the physics community: it had both the appeal of gradualism and the displeasure of seeing provability slipping away.

The ambivalence toward gradualism may be due to a need for some assurance, the way youngsters want to see the world in black-and-white, in a dichotomy. Perhaps the dichotomy makes us uneasy because we prefer to be open-minded and ready to progress, yet we still want things simple and easily manageable somehow. Thus, the dislike of gradualism rests on our dislike for paying a high price for a desired commodity. Whenever we have trouble deciding to purchase a costly item, awareness of the cause of the trouble helps but only partially. In shopping this is fairly obvious, but in other transactions less so. In political theory, the great advantage of gradualism is in its pluralism and thus in its toleration. This takes us back to the very beginning of our discourse: how do we put in practice our recognition of the siblinghood of humanity? Clinging to our background is dogmatic; allowing all regimes is irresponsible; looking for the ideal regime, namely, for the scientific utopia, is the solution that Plato sought; it does not work because it needs the absolute knowledge of the philosopher king and he is nowhere available. Thus, pluralism is the only way out; it developed as a coherent philosophy only after World War II, in the shadow of the Holocaust. A big price to pay for any lesson whatsoever. Pluralism is limited. It tolerates options but it is not averse to grading them, and to efforts to do away with the worst opinions and thus with the worst regimes. This is enlightening.

The aim of most writers on government was to render politics more humane. How to achieve this is still not clear. What is Eurocentric here is liberalism. Now not all humane writers are liberal. Western liberalism stands out as both self-conscious and grounded in the legal system. This is the central Western liberal rule that already Machiavelli has worded: try to write laws in such a way that the public will find it in their interest to abide by! The grand example is, a national militia is preferable to a mercenary army because it is in the interest of the citizens to defend their country. This, said H. G. Wells, made Machiavelli the father of modern nationalism.[8] Indeed, the modern nation state developed—gradually—as a law-abiding civil society that tends toward egalitarianism. Unfortunately, in the wake of the failure of the French Revolution some German philosophers, mainly Fichte and Hegel, did all they could to destroy the very idea of equality, by fighting the idea of individual independence, and by discussing national independence as being opposed to it. This was to become the modern versions of the hostility to individualism: the individual finds salvation in devotion to the collective. It in turn led to the idea of the supreme necessity of harmony between the individual members of the collective, harmony that the collective achieves by imposing standards on its members, namely the national character. National character may sound innocuous until one learns that it was the rationale for the Holocaust. Pluralism evolved after the Holocaust and as a "natural" response to it. No one attributes it to any single author.

A problem is lurking here. A pluralist philosophy does not render legal any illegal activity, and if it allows any activity that is legal; what then is its content? The answer is that the domain of pluralism is neither the forbidden nor the allowed but the recommended. Here another problem lurks: in all social and political set-ups, some degree of pluralism is unavoidable, and when excessive, it may generate social instability. Here, however, is where we see it in action. Normal people need to identify with their tribe. This need for identification is what Nietzsche has called the herd instinct. It overrides the autonomy of the individual; only supermen are free of it. Here we see in Nietzsche the influence of the tribalist, Romantic Hegel, who allows autonomy only to geniuses. (The genius is one who overcomes the pressure that according to Hegel normal society does and should exert on all deviants, the pressure that only a few can overcome.) The need for national identity obviously requires some national uniformity. It obviously requires some ideology or doctrine. Therefore, it comes in variations. In Germany, the general view was that Jews do not share it, that they comprise a tumor in the body national. A liberal version of the same doctrine dominated the United States, known as the melting pot. Woe betides those who did not melt. The intellectual part of this national identity doctrine is what according to the great nineteenth-century liberal scholar Ernest Renan the national political leadership should fabricate as a beautified historical narrative. Gellner called this narrative the national standard textbook and recommended that it should be taught (not as lies but) as fiction. This allows the transition from romantic to pluralist nationalism; it reduces the risk to individual autonomy that the need to identify with a collective comprises.

Pluralism attenuates the problem of social order; it does not dispel it. Hence, the need for constant watchfulness in defense of democracy and individualism. The national leadership controls the education system, and this should inculcate the loyalty that goes together with the critical attitude that is essential for the retention of individual autonomy. This doctrine is admittedly European, perhaps also Eurocentric; pluralism renders acceptable to members of diverse non-European cultures to borrow freely from Europe this or that item.

The European theory of rationality, the only rational choice the proven, is an obstacle to pluralism. The European tradition of skepticism always attenuated the traditional European idea of rationality. This attenuation is what Lessing expressed in his play, *Nathan the Wise*. This suggestion of Lessing, that doubt should generate toleration, does not sit well within European culture, as this culture advocates proof of any idea offered to the public. Kant proposed censoring the publication of conjectures; Popper went the opposite way. He suggested as the hallmark of rationality openness to criticism rather than openness to proof. The idea looks obvious until it bumps with the idea that the mark of science is scientific proof, however specified. To overcome this obstacle, one has to revise the hallmark of science. This move presents a few Eurocentric ideas—of rationality, reason, science, democracy, pluralism and more—and paradoxically makes it easier for non-Europeans to join in without the loss of their cultural identity.

My claim places Popper in a highly privileged position as a thinker of the first order. People gain such a preeminent position with little ado only within science, and hardly even there. Elsewhere they have to cultivate their own schools. Popper could not possibly do that. On the contrary, he taught that the role of schools is to advocate dogmas and to allow their modification only surreptitiously. Without a school and without proselytizing Popper's message travels slowly, from one autonomous scholar to another. What helps here is the anti-Baconian Popper-style idea that agreement and appreciation differ, that we can agree with little appreciation and appreciate ideas we do not consider true, for example as daring new solutions to some difficult problem.

The difficult problem that Popper has solved is one that the great Russell admitted he could not solve, we remember: how can one avoid both skepticism and dogmatism? The obstacle to Russell's efforts to solve this problem is the very way he posed the question. These belong to methodology—how do I learn?—and epistemology—what do I know? The reference to first person singular is a token of individualism and a recognition of the fact that only individuals, not societies have brains. There is, however, much fiction in this: when philosophers of science ask, why is Einstein a scientific authority, the question is social; philosophers of science translate this question to the question, why do I believe Einstein? They do so casually, not asking themselves whether they believe Einstein. More precisely, do we believe, say, Einstein's theory of gravity? For this, obviously, we have to know what it says. Unfortunately, most philosophers of science do not know what it says, since it is an assertion written in the language of the calculus of tensors that few philosophers of science master.

Here Einstein's theory is just one example. The traditional way of the wording the theory of knowledge in the first-person singular refers to a fictitious individual who is familiar with all received scientific doctrines. We should give up this fiction. Yet replacing the first-person singular with the first-person plural raises the question of individual autonomy. Yet the question is pressing anyhow. Traditionally, individualist social thinkers viewed society as a mere conglomerate of individuals, and collectivist social thinkers viewed individuals as fully integrated within their societies, as their members in the sense of limbs, items that are devoid of all autonomy. The reasonable alternative idea is that neither individuals fully determine their societies nor societies fully determine their individuals. This idea, however, raises a difficulty: individuals and societies may be in conflict. Early in the twentieth century, Georg Simmel advocated it and insisted that conflict is inevitable, that efforts to overlook this fact leas to utopianism, no matter how reluctantly. Yet he too could not offer a reasonable theory of conflict because he still clung to the traditional idea that science is proof: *malgré lui*, his image of science was utopian too. Popper's view of science offers a natural habitat for Simmel's theory of conflict. For those who struggle with Eurocentrism this should serve as a welcome relief.

NOTES

1. The scholastic formula for the mystic union is, making the intellect and things equal (namely, fuse).

2. Francis Bacon refers to the mystic formula in his *Novum Organum*, Book 2, §19, 2012 (1620).

3. Of the latter, the most important one is Descartes, who viewed "I think" proven and so also "I exist." Spinoza's criticism of Descartes is thus deadly: he said, the famous "I think, therefore I exist" drops out of Descartes system: it rests on the ontological proof of the existence of God.

4. A funny example is the table of chemical elements. Standard history of science texts refer to that of Mendeleev. They overlook the tables that precede his as well as those that antecede it, and the correct it tacitly. For more examples see my *Science and Its History*, 2008.

5. Thomas S. Kuhn, *The Structure of Scientific Revolutions*, 1962.

6. Joseph Agassi, *Radiation Theory and the Quantum Revolution*, 1993.

7. Thomas S. Kuhn, "Replies to My Critics," in Imre Lakatos and Alan Musgrave, eds., *Criticism and the Growth of Knowledge*, 1970.

8. Machiavelli developed his idea of national army in his *The Art of War*, 1965 (1621). Wells commented on him in his semi-autobiographical *The New Machiavelli*, 1911.

Chapter 5

Magic and Science

Eurocentrism is many things. It may be disagreeable, like "white man's burden" (which is but a euphemism for modern imperialism, even though its inventor, poet and author Rudyard Kipling, sincerely believed in it). It may be agreeable like the natural sciences and modern technology, including the information highways. We should admit exceptions, though. For example, Kipling was a deeply convinced imperialist. To convey this idea, he used his tremendous ability as the first-rate writer of fiction that he was.

One of his greatest stories is "The Man Who Would Be King" (1888). It tells a story of an ignorant, selfish British soldier who found himself an emissary of European culture, able to bring it to a remote corner. This made him transcend himself and make real efforts to rise to the occasion, an effort that in the end cost him his life.

Sir James Frazer divided human cultures into three historical stages, magic, religion and science. His aim was probably to combat religion as a remnant of primitive thought that was historically understandable but rationally dismissible in the light of contemporary science. In the same vein, he declared magic to be pseudo-science, meaning, presumably, that its ideas were once tolerable, but in the light of contemporary science (he was writing in the first quarter of the twentieth century) are to be dismissed. An interesting question arises. That not all pseudo-science is magic is obvious; so, which pseudo-science is magic and why? To begin with, what is science and what is pseudo-science?

To this Frazer had the most naïve answer: the mere claim for scientific status for a theory that does not work renders it pseudo-science. Who has to make the claim? Is success the same as truth? The affirmative answer to this question is very famous; it has a name: pragmatism. To say that if a theory is true is to say that it is successful, then, it is true to say that it is successful when it is successful. Does successful cheating of the public about the success of a theory render it true? Surely not. Why? Because the deceit will not stay successful for long. So, then, to count as scientific a theory must be successful and its success has to last. How long? And what significance is there to errors that one culture deems true and another does not? In brief, pragmatism is so foolish that one has to ask, how did such brilliant people as the American pragmatists advocate it? The answer is that they wanted to avoid dogmatism, and in particular, in the name of the siblinghood of humanity they wanted to avoid Eurocentrism. Alas, they failed, and so the question they have solved has to return to our agenda.

Take the popular (alleged) paradigm of magic: "Open Sesame!" This is (not magic but) technology. That in the story of Ali Baba the magic word was not magic proper but a password is clear from the continuation of the story: when one forgets the password, the door stays closed. Now, of course, we may declare magic the very ability of any password to open doors. This will render magical all science-based technology. Indeed, just at the opening of the scientific era, in the year 1600, Giambattista della Porta published a book by the name of *Magia Naturalis*.[1] And so, if "Open Sesame!" is magic, then any useful procedure that we consider (science based) technology, (from the use of herbs as medications to open-heart surgery) comprises a specimen of some version of magic and the impressive modern versions of it (such as open-heart surgery, may even qualify as super-magic). Alternatively, what is "real" magic?

If you want to remember how magic feels, start taking words in your mother tongue literally. A storm, then, to take a simple example, is anger; real anger; rage. You may ask, then, who is angry when a storm occurs? This question betrays your scientific bias. Who says that anger must have a subject? Magic allows attitudes to float freely in space, possibly looking for an object to cling to. We often describe this floating anger, or any other floating quality, as belonging to a spirit. In doing this we are attenuating the magic and bringing it closer to science. We speak of a nymph as the spirit of a fountain, and of hamadryad as the spirit of a tree. This distortion introduces science into magic. A nymph is a fountain and a hamadryad is a tree.[2] Take Newtonian mechanics: it has attraction and repulsion. In science these terms are merely metaphorical. Can you imagine atoms loving and hating? Can you imagine mere wishing to be a cause? As any magically minded expert will tell you, if you wish something really hard, it may really happen. Is this magic? Not necessarily. In the philosophy of Leibniz, it is not.[3] Leibniz, who was Newton's chief opponent, complained about Newton's introduction of attraction and repulsion into science. So, let us take a less controversial case. Prayer sometimes invokes magic. Maimonides said that prayer is an effort to force the Good Lord to bend to the wish of the person who prays, so that it is an insult to Him, so that prayer is objectionable. Oh, he said we are instructed to say our prayers, and so we should, but we should not mean it; rather, while praying we should think of our lives, take stock, and seek ways to improve.[4] Maimonides knew what magic is, and he found it repulsive and very disruptive, whereas religion he found conducive to cooperation and thus indispensable. Most scientifically minded religious people agree with this judgment. They make too much of a concession to magic! Indeed, Maimonides lumped all that he disapproved of under the heading of magic. He said, all idolatry is sun-worship and sun-worship is astrology; and, he added, astrology is a sort of magic. He probably went too far, but he was right in suggesting that the magic worldview is animist. He contrasted the magical worldview with what today we call naturalism—the way Hume understood this term.[5]

Without exception, all non-European cultures are magic-minded. So are all children. Parents encourage it when they encourage their children to kick a piece of furniture that they have bumped into—in order to divert their minds from the pain, of course, yet the action is animist, and so it is magical. Now science is anti-magical; yet the popularity of scientific thinking in the west did not oust magical thinking even there.

Under the influence of Malinowski, anthropologists find Eurocentrism an obstacle to the scientific objectivity that they should display. Therefore, they find most objectionable any dismissal of preliterate peoples as primitive, and so they find most objectionable the attribution to them of the belief in magic. The fictitious idea of the Noble Savage of seventeenth-century Europe gave way to the equally fictitious Europeanized preliterate peoples, all in the name of the siblinghood of humanity! The way Europeans neutralize the magic that their languages are soaked in is by reading it metaphorically; and this is what, they claim, preliterate people do too. Take animal sacrifice. We transfer our sins to the animal and punish them for our sins by slaughtering them. This is magic. We can view the rite metaphorically, and we nay abolish it. Not all who keep saying their prayers ascribe to them real power. Why can we view believers as magically minded but not the preliterate tribespeople? Because, while we speak of the devotees in our midst, we do not consider their very presence as casting doubt on the siblinghood of humanity. Hence, the social anthropologists who try to adhere to the doctrine of the siblinghood of humanity simply overreact. Whenever some people overreact, some other people wish this explained, and then the psychoanalysts say, overreaction reveals hidden attitudes to the contrary. At times, this is the case; whether it is always the case is hard to judge. The matter requires more investigation. For example, we do not quite know what is ambivalent. Konrad Lorenz, who appreciated Sigmund Freud despite his Jewish ancestry, nevertheless criticized him validly by pointing out that some cases that look like ambivalence may be no more than ambiguity. For example, Freud said, homosexuality is ubiquitous and the outcome of ambivalence; Lorenz said, these cases are often outcomes of ambiguity. The psychoanalyst Erich Fromm said, Lorenz's theory is sometimes fuzzy. And so it goes. When discussing magic, we can see both ambiguity—lack of clarity as to what magic is—and the ambivalence toward magic that prevents the proper study of it.

To repeat, unlike the magical worldview, the scientific worldview requires that every characteristic should belong to some object. And so, the wave theory of light—the idea that light is undulations—requires that whatever is between the source of the light and its receiver, the medium, must be an object: the ether; the idea that empty space undulates sounded illogical or unscientific, if you will. This made Michelson look for the ether in an 1879 experiment—for which he won the Nobel Prize. Although he died in 1931, decades after Einstein's theory won almost universal consent, he never allowed for the possibility that there is no ether—his own failure to detect it notwithstanding. It is not clear to me why. Already in 1840 Michael Faraday said, "I dismiss the ether but not the vibrations"; this led to his total isolation from his peers, despite the great fame and adulation and rewards that the world of science had formerly bestowed on him.

Ian Jarvie has noted the peculiar way in which anthropologists face magic: they consider it problematic. They observe, say, a man sowing rice in a field and then dancing around that field. They find the sowing reasonable, of course, as an empirically attested fact; but they wonder what the reason for the dancing is. It is of course magical. Reluctant to ascribe to that man the view that the dancing has any influence, they take the dance to be an empty ritual, akin to the retention of prayers in Europe that some people perform even after they have lost their faith, much the way Maimonides said prayer is compulsory but the hope that it be effective is not. Jarvie has presented their problem in a very helpful manner since it led naturally to his criticism of it. This criticism is very significant, since to remove a problem amounts to the removal of the theory that comes to answer it! Jarvie's criticism of the problem is simple: the anthropologists who see the preliterate farmer sow rice and dance are in error: he sows rice; his dance is part and parcel of his act of sowing. Proof: the very discovery that the dance is a separate act is the discovery that the expectation that sowing without dancing will be fruitful is reasonable; it is earth-shaking; it leads to the choice to retain the act of dancing as merely symbolic and thus to its possible harmless abolition.

This raises a simple problem: how do social anthropologists explain the church going of those who have not lost their faith? The answer that anthropologists give or imply is that of their grand ancestor Émile Durkheim: no one ever believes: religion is and always was an empty ritual. Proof: faith is irrational and people are rational. QED. Silly as this claim sounds, it has deep grounds in the philosophy of religion: religious and non-religious people alike agree that faith comes to supplement reason. Positivists declare faith opposed to reason; non-positivists disagree—from the religious Maimonides to the anti-religious Russell. Indeed, religious anthropologists had to protest against their positivist peers. Evans-Pritchard did. He salvaged not only his own religion but also the religion of the preliterate tribes that he had studied. He then had to ascribe rationality to their magical rites. This he did. Of course, as a religious person he had a more limited view of rationality than the positivists do: he took it for granted that the very intellectual frameworks that people live in are given rather than open to rational choice. This is a pluralist view of reason, now more popular than in the days of Evans-Pritchard. It is hard for me to see why thinkers do not see that some people are better able than others to choose their intellectual frameworks. In any case, Evans-Pritchard rightly took criticism as the mark of rationality. He observed that the preliterate people in whose midst he lived for a while in order to learn their ways regularly tested the oracles that they were consulting. Given the oracles and the reasonableness of consulting them, what remains is the choice of an oracle, and this is open to critical examination that is the option of rationality that the intellectual system of the preliterate people can and does employ. This holds for the Greeks and the preliterate people alike, and so it is not Eurocentric in any way. What is Eurocentric is ridicule of the choice of oracle: among the Greeks, it was the Pythia of Delphi; among the tribe that Evens-Pritchard has studied, it was chickens. I was present when he told this to an international group of distinguished social scientists. His mention of chickens made them laugh. He left the lecture hall with tears in his eyes. They will never believe, he told me, that I mean it seriously. This is still another kind of Eurocentrism—one that has little to say in its favor; perhaps not: perhaps it is but one facet of positivism. For, however narrow-minded I consider positivism, it seems to me that taking science for your dogma is the best dogmatism around. Although positivists are not as utterly free of superstition as they boast, science does free them of many dangerous superstitions. Interestingly, (1) positivism and (2) new-age spiritualism are the two unintelligent Eurocentric radical versions of (1) anti-magic and (2) magic (European style). They survive, despite their folly, because (1) magic is a part of life and of art, and (2) it is a folly that looks very intelligent. My paradigm case here is the output of Carnap, one of the leading European positivists of the twentieth century. He found it most important to explain the meanings of words while rejecting the claim of his teacher Frege, the father of modern logic, that meanings dwell outside space and time, namely in the domain that we call Platonic Heaven. He was not the only one engaged in this activity. Russell, Frege's leading follower, tried in his most famous 1905 paper, "On Denotation," to show that there is no need for it. He continued in this effort until 1940. He then admitted defeat whilst sticking to his opinion nonetheless. Carnap declared repeatedly that he had reached the target: it is now possible to dismiss Plato's Heaven. Toward the end of his life, he expressed optimism: the aim is nearly reached. Meanwhile, the leading logician of the next generation, his admirer W.V.O. Quine, defied positivism. He told me he gave up hope of receiving a fair review of any of his books, since no reviewer would allow for his—Quine's—view that the number 1 exists like tables and chairs. Later on, the famous logician Mario Bunge came up with what is the leading alternative today. It is that we should allow for Plato's Heaven, but as a mere piece of fiction. Ingenious. My alternative idea is that there are diverse kinds of existence but this proposal is too radical for leading logicians to take seriously.

An interesting, significant historical fact is that both magic and anti-magic played a major role in the rise of early modern science. To see this, we need a general view of European culture such as Harry A. Wolfson the great twentieth-century scholar provided.[6] The most conspicuous aspect of it is that (unlike ancient Greek culture) it was primarily religious. The three Western religions, says Wolfson, are variants of one: monotheism. They interacted and influenced each other. Yet this single religion had two versions, rationalist and mystical. The mystical idea, incidentally, never left the rationalist camp because rationalism was the (false) idea that the rational is the provable, and there was always only one theory of proof, namely that one acquires knowledge by the mystic experience or the mystic union. This theory itself has two versions, irrationalism and rationalism. The irrationalist version says that knowledge is ineffable (not given to verbal expression). The rationalist version, says Wolfson, landed in the mediaeval university. The mystics were loners, usually hermits; when they entered society, it was in courts: they were thus mainly courtiers. Until the rise of the Royal Society (1660 or a bit later), all revolutionary thinkers of the Renaissance were courtiers with the exception of Copernicus and Galileo whose affiliations were unusual, while the universities were hostile to science until the French Revolution, at least. They really became less hostile to science only when Humboldt, both a thinker and a politician, legislated that universities must be enlightened. More realistically, only much later they became pro-science and no questions asked.

To repeat, Copernicus and Galileo were neither academics nor courtiers. Copernicus had a Church position although he was no clergyman; Galileo was most unusually a courtier affiliated with a university. They were both, however, clearly opposed to the academic tradition, and thus, of necessity, mystics. Thus, Copernicus said that the sun is the symbol for the throne and the place for the throne is always in the center of the court. The word for mysticism, incidentally, was cabbalism, which means tradition, meaning verbal tradition, meaning a secret doctrine transmitted from master to disciple. The source of the tradition is of course Moses the lawgiver. As the fifteenth-century mystic Regiomontanus reports, it was when Moses came down from Mount Sinai (allegedly eleventh century BC) he met with Pythagoras (sixth century BC) and told him all the secrets. Since the word

"Cabbala" is Hebrew, it was altered to "Pythagoreanism."[7] The opening of Galileo's mind-boggling *Dialogue on the Two World Systems* is most impressive. The debate is between an Aristotelian and a Copernican. The Aristotelian opens: you say we are confusing, but you are worse. The Copernican admits the charge and promises to behave. This is how Galileo inaugurated the era of modern science: by raising the standard of clarity. For that, there is the need to oust defensiveness. Galileo admired Copernicus, as he was not defensive and admitted that he had no answer to some questions. (In particular questions to do with the flight of birds, that later Galileo explained by the law of inertia. Before that Copernicus had said (preface of his great book) that since the ancients disagreed among themselves, they cannot serve as authorities.

That parties that disagree cannot serve as authorities is obvious. As Bacon has put it, disagreement is evidence of ignorance. What thinkers since Philo Judaeus did about disagreement—until Benedict Spinoza put an end to it, says Wolfson—was to reconcile all disputes.[8] The historically most important example is the conflict between Aristotle and Archimedes on floating bodies. St. Thomas Aquinas reconciled them. It remained to be rediscovered that they are in conflict. The Aristotelian Francesco Buonamici achieved this. Galileo dissented from him respectfully. Indeed, although few commentators notice this, it is the most important transition between ideas of the middle ages and those of the Renaissance. In the Middle Ages, contradictions were resolved. That contradictions are resolvable, and even with ease, is ancient knowledge. Resolving contradictions is easy and usually very boring. The Renaissance, by contrast, sharpened disagreements. In particular, it preferred the mystic tradition to the Aristotelian tradition.

The very idea of respectful dissent is a tremendous innovation. Many still cannot respect false valuable ideas. How are we to respect a false idea without showing disrespect for the truth? These are good questions, yet tradition deems them puzzling rather than intriguing. This should not be so.

Galileo, and before him the ancients—especially Democritus—developed the most radical anti-magic idea of causation. In the modern world Galileo, and following him closely Descartes, developed the mechanical philosophy largely as a barrier against magic. Whereas in magic action can be at a distance, Galileo and Descartes said, in science all actions are by contact. This idea condemns Newton's mechanics. Of course, Newton himself was aware of it; he tried hard to reconcile his theory with the Galileo-Descartes idea. So did many followers of Newton, including the great Laplace after the French Revolution. They all failed, and already a generation before Laplace, Boscovich proved that all such efforts are doomed to failure.

One of the major tasks that Immanuel Kant undertook was to legitimize Newtonian mechanics by declaring action-at-a-distance admissible. Einstein abolished action at a distance for good, yet at the cost of reintroducing another magical characteristic: a property with no carrier. What has remained of magic is intention with no carrier! Good luck is a good example: it hangs about and we wish hard that it should land on our shoulders! So good luck to you!

NOTES

1. English version, *Natural Magick,* 1659.
2. More generally, any "dead" metaphor taken literally (deemed "live"), enchants the world. See I. C. Jarvie and J. Agassi, "The Rationality of Magic," *British Journal of Sociology*, 18, 1967, 55–74. Reprinted in B. Wilson, ed., *Rationality*, 1970, 172–93; in Steve Bruce, ed., *The Sociology of Religion*, Vol. 1, 148–167; in J. Agassi and I. C. Jarvie, eds., *Rationality: The Critical View*, 1987.
3. Joseph Agassi, "Leibniz's Place in the History of Physics," *Journal of the History of Ideas*, 30, 1969, 331–344.
4. Joseph Agassi, "Reason Within the Limits of Religion Alone: The Case of Maimonides," in Yoav Ariel, Shlomo Biderman, and Ornan Rotem, eds., *Relativism and Beyond, Philosophy and Religion: A Comparative Yearbook*, 1998, 125–177.
5. Joseph Agassi, "The Unity of Hume's Thought," *Hume Studies*, 10, 1985, Supplement, 87–109.
6. E.g., Harry A. Wolfson, *Philo: Foundations of Religious Philosophy in Judaism, Christianity, and Islam*, 1947.
7. Moshe Idel, "Johannes Reuchlin: Kabbalah, Pythagorean Philosophy and Modern Scholarship," in *Representing God*, 2014, 123–148.
8. Harry A. Wolfson, *The Philosophy of Spinoza: Unfolding the Latent Process of His Reasoning*, 1934.

PART II

The Revolt against Reason

Chapter 6

The Problem of Rationality

This chapter is about the problem of rationality alone: the problem itself, not any solutions to it, not even the extant proposed ones: these will hopefully come up in the subsequent chapters. Admittedly, it is unusual to discuss problems without offering some solutions to them, at least some extant solutions. It is not obvious to me why. A problem well put, a popular saying goes, is half a solution. Hence, a book, say, a philosophy of science text devoted to the problem of induction—I will present it later—is better written if half of it is devoted to the problem: what is it and to whom is it important and why. Every year quite a few papers devoted to the problem appear and every few years a book on it appears too. Often, publishers' brochures herald them as path breaking. So do reviews of them and other sorts of comments that appear in the learned press. Yet hardly any one of those states the problem and almost none discusses it.[1] Popper, in his first book, used a great deal of space to discussing the problems of the philosophy of science, their formulation, and the obstacles to whole swathes of proposed solutions.

Nor is it easy to state the problem of rationality. Quite a few versions of the problem of rationality appear in the traditional philosophical literature, and each version appears in a few variants. Many texts that discuss the problem do not pose it. Indeed, in my experience I seldom found a philosophical text that states a problem. More often, an author does not even refer to a problem. (This is fine if the work is narrative; not otherwise.) Quite a few authors refer to a problem without stating it; others even discuss its background at some length, and perhaps its extant tentative solutions, yet without posing it explicitly. This is a common menace. It seems so simple: start with an interrogative word and end with a question mark. But no. Why is it not simple? Because of so many obstacles. You may pose an insignificant problem, and then your text is an invitation to waste time. Alternatively, you may pose a valuable problem, and then your text is highly pretentious. Who do you think you are?

Who do you think you are? I have heard this question many a time: colleagues love to throw it at students. I have never seen it in print, though. And rightly so: it is unforgivably offensive. Hence, we should never ask it; if we can help it, we should not be close associates of ones who do. Just think of it: if you publish a light political pamphlet, then you may already surpass yourself. Naturally, you hope to have some impact on political life. Even if you do not write a pamphlet but merely peddle one, if you perform any political action whatsoever, you thereby already express the hope of making some difference, and thereby you expose yourself as megalomaniac. Who do you think you are? You are a nobody, and you pose as a person able to influence others politically! We should expose you as a windbag. The outcome of this policy is support for the status quo, for the current political leadership. This support may be commendable, but not as a matter of course. Hence, the right answer to it is, never mind who I am; the question is, who should be our political representative.

Suppose you manage to be original or have access to some original idea that is still new. Before discussing it we may ask, how does such a strange event come about? Most cultures have no new ideas, or so few of them that they raise no problem. In the nineteenth century Lewis Henry Morgan asked, how is it that so many societies lag behind Europe? His discussion of this question is outdated, important though it once was. The strange thing is that today we ask the question the other way around: how did progress happen in the first place? Carnap, one of the leading philosophers of the mid-twentieth century, took it for granted that all reasonable people can come to full agreement on any question if they come together in good faith and try to settle their disagreements. This seems to me nothing short of amazing. It did not occur to him that to reach total agreement might mean total stagnation. Suppose you manage to be original or have access to some original idea that is still new. As you express it your audience will ask, can you prove your idea? Are you sure you are right? Can you convince us? Do not answer such questions. They are traps.

There is a general point here, too significant to bypass. A question, said the great pragmatist philosopher C. S. Peirce, is an invitation, a request, to choose one option out of a list. Sylvain Bromberger accepted that idea and found it surprising that we understand "why" questions even when we have no candidate for an answer. We know that we understand the question, since when we learn of an answer, we know whether it is adequate. Indeed, it was the great insight of Peirce that before trying to examine whether a given answer is true, we should examine whether it is adequate. And, of course, to ask a question is an invitation to choose one out of a list of adequate ones. How is it then that facing a why question with no candidate for an answer, and we face then a new answer, we know that the answer is adequate? Take a real and important example. Say, why is the sky blue? or, why is the grass green? or, why is the sea emerald? As it happens, until 1900 no one had an answer to any of these.

The first one was answered first, as the answer belongs to classical physics. The other two questions happen to be harder and answered by later physical theories. No one could imagine that this is how things would turn out. Yet even children easily understand the questions.

How come? What makes a question understandable? Why does familiarity with science improve our understanding?[2] These difficult questions have led the leading logician Quine to a holistic view of language, from which it follows that no two people mean exactly the same thing by any word or sentence. So even when two people agree about the range of a question what they share is not quite the same question. Yet they usually understand each other well enough. That a conversation includes a lot of background material becomes obvious by paying attention to the need for training in order to acquire the ability to read centuries-old stories. Notice that coming to understand questions can be hard to do renders the Bromberger puzzle less puzzling.

In social affairs, we often face a problem that pertains to the adequacy of an action to a situation, including the adequacy of an answer to a given question in a given situation. The question, who do you think you are, may be reasonable—not nicely put but reasonable nonetheless—say, when asked in some formal context. To take the paradigm case, if you say, I now pronounce you etc., we can direct this question to you to mean, are you authorized to do that? Admittedly, in such a context the question is unproblematic, even if its wording happens to be insolent. In other kinds of context, there is no adequate response to this challenge. You should then refuse to accept it since in that case you are bound to lose. You can accept the challenge and make good on it only if you are already a recognized somebody. In which case no one is likely to challenge you in the first place. Challenge comes in many ways. Suppose you do not act politically and you are careful not to make any demands on others. Suppose you merely make an assertion. If it is insignificant, then you are wasting our time; if it is significant, why are you of all people the person who asserts it? Worse: suppose you are in a privileged position to make an assertion, say because you are an official, or because you have researched the area within which your assertion falls. You may then face a challenge: prove it! The simplest way to counter such a challenge, says Bernard Shaw, is to issue a counterchallenge: disprove me! The acceptance of the counterchallenge is wonderful; except that if you are a student and you argue with a professor, you will not dare to pose a counterchallenge—in the hope to become a professor someday or otherwise become a somebody someday. How does one become a somebody? If you think that you are a somebody, you thereby invite a challenge: who do you think you are? Thus, there is either an easy way or no way to meet the challenge. When someone asks you, who do you think you are, the right response is to ignore it. Otherwise, it is to turn around and walk away. At times, you cannot. For example, in a job interview not answering a question properly may be too expensive. For another example, in law courts the evasion of a question may lead to a penalty for contempt of court. This is why in court you better have a lawyer to defend you! There is no escape from it: for a discussion to be fair there should be no rank pulling. If there is, then the discussion may be a sheer waste of time: if rank-pulling is adequate, then the winner will be the one who pulls rank, except in cases in which defiance may meet with reasonable success.

A fair discussion should be impersonal or disinterested from the very start. Rank-pulling then is a mere distraction. Or rather, it should be, and in rational discussion it does not come up and if it does it is shot down fast. Otherwise, it is best to consider it tacit defeat: I cannot concede defeat explicitly because of my rank. To see this, we may need to know the difference between the disinterested and the uninterested. At times, this difference is very important. The uninterested in a discussion should not participate in it; the interested in a discussion tends to participate in it the wrong way, by displaying a bias. So, we should be disinterested; except that we do not know how to and we do not know if we succeed in behaving so.

There is no way to get out of this trouble: to conduct a useful discussion we must be indifferent to personal dispositions. The young, almost anonymous social worker and schoolteacher Popper attacked the central doctrine of a world-famous group of philosophers, the Vienna Circle,[3] whose leader was the physicist and philosopher, Moritz Schlick of the University of Vienna.[4] Schlick accused Popper of self-aggrandizement. Of course, Popper never answered. Carnap, second in command after Schlick, did worse: he ascribed to Popper the doctrine of the Circle that Popper was criticizing. It is like arguing with Protestants while considering them Roman Catholics. Now Carnap did not mean to be insolent: too unimaginative for that, he was blind to his own misconduct. As evidence let me mention his response to Russell's criticism of the first book of Wittgenstein. The criticism was deadly. Carnap admitted that, yet he clung to the central idea of Wittgenstein (that an assertion has meaning if and only if it is scientific in the sense of open to empirical decision as to its truth value) that Russell was dismissing. We can apply all this to the normal posing of the problem of rationality, or even to the wrong posing of problems in general: it is sad that writers do not state explicitly the problems that they discuss, but it is not immoral, much less illegal. If readers can supply the question, then the damage is minimal. It turns out to be significant if a critique of a book is published. Its author may say, I am misunderstood: I addressed not the question ascribed to me but another. It is thus better to forego reviewing a book in which such an ascription plays a serious role. When does it?

This is hard to judge, but one thing is obvious: a book is harder to judge when its author's intent is not explicit yet the judgment of the book depends on it. The intent can be humble and it can be bold. The scientific tradition encourages authors to write scientific papers with humble intents and close them with a brief coda that expresses a high boast. At least a scientific paper has to state some scientific information explicitly. Authors can avoid this too. It is then not clear what they say. Popper has censured both Karl Marx and Freud, calling their ideas pseudoscientific. Which ideas? What did Freud say? What is his doctrine? It is hard to say. Let me take an example that seems to me fair. Freud's *Psychopathology of Everyday Life* (1901) has a discussion of slips of the pen. He claims there that they reveal some dispositions that speakers have but are reluctant to utter openly. He offers many examples. He saw a woman sign erroneously with her maiden name, raising his suspicion that she had marital problems. Indeed, he added, in a few months she was divorced. Now the case was not very common. The common is the hint that a slip of the tongue or of the pen betrays some message that itches to come out but is censured. Possibly this theory is true. Yet a slip that does not do that is possible, as is a censured expression that fails to come out. What then is Freud's idea? Can it be stated in a universal statement that has no exceptions? I do not know.

Perhaps it should not bother me that most texts on the problem of rationality refuse to pose it explicitly, with the use of an interrogative formulation and a question mark. I should state it here and now. It turns out to be not that simple. The problem of rationality has variants. The one that appeals to me most is the one that requires no philosophy to comprehend or to see that it is of central importance in any individual's life: what way of life should I choose? Let me admit at once, that many philosophers sneer at my wording. They say, this is a nice problem, and my concern with it is nice too, but this is not the problem at hand. Let me postpone discussion of this matter. At least I have posed a good question. So, let us proceed with it.

What way of life should I choose? This is a big question. It is easy to break it up to smaller ones. What profession should I choose? Should I be a criminal or a blue-collar or a white-collar or . . . ? Should I be religious or not? Should I marry or remain a bachelor? And so on and so forth.

One may decide on such questions early in life or move along with no decision at all. Thus, many people find themselves having made significant life decisions casually, without deliberation. Most people make life decisions in steps: one may enroll in a medical school without knowing what kind of medical profession one will choose after graduation, if at all. Is that wise? Is this the open-mindedness that welcomes opportunities or the indecision that lets opportunity pass by? How do we adjudicate such matters? In traditional societies, young people hardly know their future spouses before their wedding day; in the modern western-style world, couples enter marriage as gradually as they wish. Which custom is the wiser? Are people rational? If so, how do they solve the problem of rationality in order to be able to act rationally? This question seems obvious. It is not. To repeat, people enjoy art with no knowledge of aesthetics. They even eat before they have any ideas about nutrition. Some religious leaders use this fact to fend off any theological question that they fail to answer.

This seems to be the wise starting point for handling any question: the received practice—if there is any—is usually the default option: as a matter of fact, people take it to be so. Only sophisticated people even think of questioning received practices and opinions. Let us then take this option as the default one, no matter how tentatively. Now, as most people do not question received practices and opinions, they have no reason to face the problem of rationality. The very facing of it is for many an earthquake: often it is the outcome of a major discovery, the discovery of the siblinghood of humanity no less. This is odd, since the awareness of the siblinghood of humanity is ubiquitous. Yet we tend to forget it even though for our own peace of mind we should remember it. Most societies have not paid much attention to it; this holds strangely even for members of societies that have taken it seriously, say as a constitutional demand from legislators not to violate it. Nevertheless, most people have no reason to think about it, and so they ignore it. This has tremendous consequences for most individuals: it means that throughout history most people around the world have not made their life decisions; that they have found themselves in positions where their central decisions were made for them. The way these decisions are made then is either by tradition or by people in authority—for example, parents marry off their offspring. Of course, by default, authority is what tradition says it is; the idea of the siblinghood of humanity allows individuals to question this. One way or another, then, most people have their central life decisions made by tradition. Tradition is then the first casualty of the rise of the doctrine of the siblinghood of humanity; the waning of tradition then raises the problem of rationality. It is not surprising then that both are characteristic of the Eurocentric philosophy. As a non-European society adopts the doctrine of the siblinghood of humanity, it and its members undergo the crisis of having to make serious decisions that even in the west are often too difficult to handle properly. This may be avoidable in principle, but no political thinker has yet offered a technique for doing so. Moreover, the first cracks in tradition raise the possibility of giving up tradition as a whole, the philosophy known as radicalism. This explains the bitterness that some traditional thinkers display toward westernization or modernization or opening up to foreign influences or to reason. The cause of bitterness is pain; admittedly, bitterness does not relieve any pain, but it is no use telling bitter people this; on the contrary, it intensifies their pain and their bitterness alike.

Not that Europeans are immune to the pain and bitterness of facing the problem of rationality or of the radicalism that often appears as the first outcome of its discovery. The Western verbal arts are full of examples of this kind, of cases, of complaints of people who have the problem of rationality thrust on them with no adequate preparation. Some of these cases are tragic, other are entertaining. All are unfair. Life is not fair. A leading twentieth-century philosopher, Ludwig Wittgenstein, never discussed in writing the problem of rationality.[5] He did make it known, nonetheless, that he was not enamored with the modern world, that he was not convinced in the least that what goes under the name of progress is indeed progress. He added that he found modern life unfair and unpleasant and thrust on people unprepared. Yet he hardly discussed all this. It is no accident that the place for this discussion is the verbal arts, including short poems and bulky novels, including the cinema and the television and more.

Fair or not, it is still imperative that a reasonable discussion should be impersonal. Of course, it should discuss a question that matters—personally to participants in it as well as to the readers of reports on it—because it concerns some needs that they have, or it offers useful tools for handling these. For a tool to be useful, it should not depend on the individual intended to use it. When it does, then its details are of no public interest. Therefore, the question, are you qualified to discuss it is irrelevant. Discuss it and we will hopefully judge your performance on its merits afterwards. Perhaps after an unjust delay, if ever.

Any such claim that sooner or later all the debates that deserve public attention will receive it is not true. It is ideal, and the Europeans approach it better than any other culture, especially within the scientific community, but it still is not the fact of the matter. Moreover, the injustices even within the scientific community are at times short-lived, at times not. It is when those who were victims of injustices retire or die, and their heirs have the opportunity to rectify the verdict, that they may do so. This is not good enough, of course, but nevertheless it is better than in most parts of human society. This—science and the scientific impartiality that is associated with it—is the force of Eurocentrism. It is the reason for the material success of Europe. This is why it is all right for me to be Eurocentric—although I may not qualify, as I was not born in Europe.

Although we have not discussed solutions to the problem of rationality, we have advanced toward some possibly adequate solutions. Notice that we can advance slowly from no solution to the true or satisfactory solutions in steps. Before we have any solution to a problem, we may know what a reasonable solution is not. For example, every why question has the answer, because it is God's will. This answer is incontestable. Some people who reject all religion may protest that this answer is not acceptable. Their mistake. No matter what your religious or theological views are, you can always take the answer as a metaphor, and thus you can always reword the answer to suit your specific views. You cannot make the answer adequate, however: such answers are to be disregarded as unsuitable, be they true or false. What we require of an answer to consider it a serious candidate for a satisfactory answer are *adequacy conditions*.[6] The adequacy condition that we have discussed thus far is, disregard answers that depend on their sources: in the first instance it does not matter who initiated a putative answer: the reasons for taking an answer seriously cannot be that its source is important: adequacy should be independent of origin. We take an originator of answers seriously only after we are satisfied that the answers are adequate, perhaps also true.

Decades ago, it became clear that publication pressure is here to stay: to become an academic—or for one without tenure to stay an academic—one should have papers accepted for publication in a peer-reviewed periodical. In competition for having a paper published there, leading sociologist Robert Merton has observed, it helps to have some papers published there already. He called this "the Matthew effect."[7] This became the received opinion with no examination and no discussion, because its author was a leading sociologist whose output was published in leading scientific periodicals. To the extent that Merton's principle holds, its application is in violation of the adequacy condition just mentioned. Except that the publication of an idea in the learned press does not make it adequate. What does? Public opinion. How does one find what public opinion is? How is it possible to influence it? This is the sixty-four-dollar question. Gellner, a leading philosopher and anthropologist, has observed this: there are two sets of public opinion: of the heteronomous and the autonomous; they function as the general and the *avant-garde*. Any path-breaking idea that a recognized leader invents may win quick general recognition; otherwise, the *Avant-Garde* may consider it and then it becomes received after the *avant-garde* reports to the general leadership that it is deserves attention. This takes time.[8] During that time, the author of the new idea may drop out of the race for recognition. This is one of the injustices that the academic world takes in stride as inevitable casualties. Their mistake.

All such discussion is but feet shuffling: the great divide within philosophy is between the advocates of rationalism and the advocates of irrationalism—if we may honor those who advocate disrespect for reason with the honorific title of philosophers to begin with. I scarcely understand them: they resemble those who advocate the avoidance of the use of one's right hand: one may advocate it for the purpose of training one's left hand, or in some situations it may be important for the right hand to be free for some expected emergency, and so on. Still, to recommend it on principle sounds crazy. Yet many philosophers do recommend abstention from thinking and many others offer them criticism although it is obviously undeserved. This debate is worthless. Some philosophers observe that some problems are not open to rational solutions. This leads other philosophers to discuss the limits of reason. This is interesting but not to the point. Suppose we forbid jumping twenty feet high and justify that by observing that no one can jump twenty feet high! Some philosophers do argue this way from time to time. Thus, St. Thomas Aquinas recommended marriage as it is natural.[9] This is like recommending food. Incidentally, St. Thomas himself ate food but he did not marry. Similarly, Wittgenstein forbade asking unanswerable questions, yet the problem of induction never ceased troubling him.[10]

Who did he think he was? A Moses? No. He thought he was a Galileo, or a Kant. Why? What was his contribution to human welfare or to human knowledge? Answer: he declared metaphysicians slum landlords and declared he would not allow them to act freely. How his diktat was to be imposed, he did not say.

Probably by shaming. He forgot that Galileo had never demanded anything of anyone. When he was censured for speaking against his peers, he said they had attacked him. As to Wittgenstein, almost no one dared stand against him during his lifetime. Popper stood against him for a few minutes and he left the room. Later (in a private letter) he called Popper a donkey. Ah, well.

NOTES

1. Bertrand Russell and Karl Popper are the exceptions. They have both stated the problem clearly. Russell has devoted a chapter to it in his path-breaking *The Problems of Philosophy*, 1912. The book was path-breaking since it is devoted to problems, scarcely to solutions. He looked for solutions and discussed this search of his during the rest of his intellectual life that lasted decades and was amazingly fruitful.

2. Bertrand Russell, "The Cult of 'Common Usage,'" 1953.

3. Karl R. Popper, *Logik der Forschung*, 1935; *The Logic of Scientific Discovery*, 1959.

4. B. F. McGuinness, *Moritz Schlick*, 1985.

5. Joseph Agassi, *Ludwig Wittgenstein's Philosophical Investigations: An Attempt at a Critical Rationalist Appraisal*, 2018.

6. Dov M. Gabbay & John Woods, *A Practical Logic of Cognitive Systems*, 2003.

7. "To them that hath shall be given, from them that hath not shall be taken away" (*Matthew* 13:11).

8. Joseph Agassi, "Cultural Lag in Science," in J. Agassi (Ed.), *Science and Society: Studies in the Sociology of Science*, 1981, 119–131.

9. St. Thomas Aquinas, *Summa Theologiae*, Supplement, Question 49 (2006).

10. The last work that Wittgenstein wrote was his longest, on the problem of induction, that is a dud. Of course.

Chapter 7

Theories of Rationality

The use of reason comes naturally. One hardly needs philosophical training to act rationally or to think logically. One needs less any awareness of one's rationality or of one's use of logic. Even other animals use their reason freely. As the ancient stoic philosophers noticed, even animals use logic. Proof: when a dog follows a trail and comes to a fork, having checked all options but one, the dog follows that one option with no test. There are also more direct manifestations of animal intelligence. For example, cats and dogs can open doors not only by pushing them but also by pressing handles. The division of labor between insects and between pairs of birds is quite sophisticated. Foxes can deceive those who track them by moving back on their tracks. It is equally obvious to all that we humans are much more intelligent than other animals—since we use abstract language. The idea thus appeared repeatedly in the modern world since the eighteenth century that teaching apes to speak should raise their thinking ability. This some animal psychologists have tried. The way to understand the outcome of their experiences on this question, however, is under debate, since it is not clear whether moneys that use simplified keyboards are aware of the processes that they undergo sufficiently to call it a use of abstract language.

All this has led naturally to the received view that the concoction counts, of intelligence, wisdom, rationality and knowledge. This found its expression in religious homilies of folk wisdom—expressed in parables, fables, allegories and tales. This concoction is tradition or commonsense or folk wisdom or something of that ilk. This becomes amply clear in disputes. Thus, some folk sayings ascribe to women or to black people abilities that other folk sayings claim that they lack. These days this kind of discourse goes by the new name of folk science. And then, the comparison between our folk science and yours comes up naturally, just as the taking it for granted that our folk science is obviously the best. Yet to this day scientists dispute ascriptions to women or to blacks of some talents or their absence, thus proving that we can still find all sorts of pseudo-science in our highest intellectual echelons.

Perhaps this is too rash. The idea of pseudo is vague: when we declare any item pseudo or affected, bogus, contrived, fake, feigned, mock, phony, pretended, sham and their likes, we suggest some deceit: we do not call pseudo any substitute for an item, any mock item. The celebrated mock turtle soup is not a pseudo turtle soup but a decent substitute. Hence, when we call an item pseudo we imply or even allege the presence a false pretense, we accuse someone of intendedly misleading. When we consider any theory pseudo-scientific, then, we thereby make an accusation. Now it is obviously impossible to accuse a theory; the pretense is in a claim (that the theory in question has scientific status); it is possible to accuse those who make the claim for a theory for that status that it does not have, be it the status of science or any other status. Thus, it is quite impossible to view a biblical idea as pseudo-scientific, even though science has superseded it, and even though quite a number of present-day religious thinkers are. Still, some scientists seem sincerely religious, and it is unfair to declare them insincere or bigots without some clear evidence. Likewise, some advocates of science also follow traditional, established religion and it is at times hard to show them inconsistent. And then the question arises, what makes for the status of science and what does it amount to. The first question has a traditional answer; by the meaning of the word "science" (*epistēmē*; scientia): a theory acquires scientific status when and only when it is proven. The claim that it is scientific, then, is the claim that it is proven, be this claim true or false, honest or deceptive. How then can one endorse both the claims that belong to the official doctrine of an established religion and up-to-date science, when they are in conflict? Obviously, at least one side has to concede. In the Judeo-Christian tradition, science always wins. How religion concedes to science is under debate, but not that religion concedes. Moreover, medieval tradition included scientific discussions of the way religion concedes to science. This makes sense in a religious world where all professors belonged to the clergy; it does not suit the secular modern world. In it the ruling holds, the proposal of Kepler and Galileo holds that the representatives of science, particularly scientific researchers, need not trouble themselves with such questions: they belong to the clergy. Whatever science is, ever since Copernicus ruled (in the preface to his revolutionary *On the Revolutions of the Heavenly Spheres* [1995 (1543)]) that science does not rest on any authority. To be authoritative, then, the followers of Copernicus required that it be proven. How?

We do not know: it much depends on our idea of what proof is or, more precisely, on our idea of what proof-procedure is adequate. How then is a theory proven? What is the evidence that it has been proven? Given that a statement is proven, how do we check this assertion?

The paradigm of proven assertions was Euclid's *Elements*, the all-time European best seller second only to Holy Writ. Theorems there are proven by deduction from first principles or axioms and definitions. Hence, the allegation that they are proven rests on the allegation that these axioms and definitions are proven. They are not. (Definitions were either verbal, namely abbreviations that are uninformative, or essential; modern logic deems essential definitions axioms.) What grants axioms their alleged special status is their alleged self-evidence. What makes a statement self-evident is unclear, but we know what deprives a statement of its status as an axiom: its being questioned. For, this proves that it is open to question, namely, unproved.

Euclid's *Elements* looks reasonable, since it is possible to dismiss anyone who dares question any thesis in it. Hence, it seems, as long as no person questioned its axioms, they maintained their status. Not so. It is always, always possible to find some fool who will question any thesis whatsoever, even that black is black or that white is white. (The German philosopher Hegel did just this, and in all seriousness. This made his near contemporary, the great William Whewell, declare him and his fans utterly and irretrievably unserious and thus people to ignore. In the twentieth century Popper said the same much more thoroughly.) Thus, we allow only serious or knowledgeable people question Euclid's axioms. And, yes, of course, one who questions Euclid's axioms is not serious. And then the philosopher Kant corrected this assertion; he said, of course, everyone can question Euclid's axioms, since otherwise it is impossible to check the claim that indeed they are certain. Rather, he said, trying to offer an alternative to them is bound to end up as a failure. This somehow angered Kant's younger contemporary, Johann Carl Friedrich Gauss, known as the prince of mathematics, and so he dissented from Kant. A century later, the great Russell, a father of modern logic, observed that all the great advances in mathematics of the nineteenth century came as acts of defiance of Kant. That era ended with the invention of meta-mathematics, the study of such questions as what mathematics is, what mathematical proof is, what is its force and what is the origin of that force. After both Einstein and Russell managed to kill the idea of scientific proof as such—claiming (both in 1905, although arriving at that conclusion independently of each other and by totally different routes) that a statement that is provable is not about reality and a statement about reality is not provable. And then things settled down. Since by now it is generally accepted that proofs are not about reality, their status fell considerably and they are not sought that avidly. We forget that once people sought avidly the proof of the existence of God. When Kant said that it is plausible but not proven, the status of theology diminished considerably: the inability to prove that God exists was deemed a disappointing collapse of reason. Even the great logician Russell was excited when he once thought that the ontological proof of the existence of God may be valid. This proof says, as God is by definition the best and all good, he is better if he exists and so by definition he does exist. Later Russell said that by the same argument the devil exists too. Yet today the wish to prove God's existence is expressed much less frequently than in the Age of Reason, since the very idea of proof has lost its force. Nevertheless, the very idea of scientific proof has not fully lost its allure. What is it?

Bacon said, the claim for a proof is very deceptive. For, once you claim that a theory is proven, you can fit to it some proof or another, and even with ease. Moreover, there may remain at the very least the possibility of dismissing some alleged proofs as invalid. Even that is questionable, he said: people who endorse a proof will dismiss with ease the dismissal of that alleged proof. Such debate, he observed, went on for centuries and settled nothing. They are all no good: it is important to block such debates as they are essentially futile. A theory then has to be scientific and thus quite uncontroversial at its very birth: the way it comes up should follow a proof procedure and thus be proven the moment it appears. This, says Popper, is proof by pedigree: the mere knowledge that a horse comes from a good stable is already the approval of it. The stable in question, Bacon added, is the laboratory, the place where information is amassed. A theory has to grow naturally out of a mass of information; it should not be forced. Research is much diligence and more patience. This is called the inductive process. And so, the very refutation of a theory is conclusive proof of its originator's violation of etiquette. The reason for the violation is, of course, vanity, the quest for fame. More Bacon did not say. When all is said and done, he admitted nonchalantly, knowledge emerges out of a mystic experience.[1] How do we check whether a theory is true or fake, then? Easy: a true theory is obvious. Hence, Euclid's axioms are obvious and a paradigm of (proven) science. The claim that the axiom of the parallels is not obvious, then, was at the time a very serious criticism.

The ancient idea of rationality, then, was the idea of science: the rational is the same as the scientific. And science (*epistēmē*; *scientia*), to repeat, is absolute and unquestionable and irrefutable knowledge: it is what has successfully undergone the process of proof.

There was no clearer idea of proof in antiquity. There was no clearer idea of proof in modern times either. What is beyond question, however, is that a proven idea is true; hence, it is not possible to refute it. Hence In the nineteenth century, well before the 1905 Einstein revolution in physics and Russell revolution in logic a new challenge appeared to Euclidean geometry: Gauss devised a test of the claim that the sum of angles of a triangle is 180° by planning to measure a huge triangle in the sky. The experiment was not well designed, since he had no idea about the precision of the measurement required to disprove Euclid.

Meanwhile, a general opinion of experts evolved: Newtonian mechanics is scientific, namely proven. The exposure of this idea as false may easily be construed as its exposure as pseudo. Such a conclusion is simply intolerable to many. And so, to this very day, many physicists are enraged when they hear historians and philosopher of science declaring Newtonian mechanics false, even though they know that Newton has postulated action at a distance that Einstein has declared impossible. For, they hear it said what is hardly possible to say: that Einstein has exposed Newtonian mechanics as pseudo, as a fraud. This, to repeat, they find just unthinkable. Perhaps this is why the great, friendly Einstein did say, "Newton, forgive me!" even though of all scientific researchers of all times he considered Newton the greatest.

So, here confusion still reigns. Odd, since the confusion is easy to avoid. It is, indeed, a very general point, usually ascribed to English common law: it exonerates any physician who practices magic if this is the practice accepted in the community of that physician. This stands to reason. Science always exhibits agreement over many facts. Some of them concern haram—to individuals and to society. People responsible for having caused them were not always accused of improper conduct. Now the excuse that one has caused harm in ignorance is no excuse. Yet the excuse is sufficient when evidence show that the people who caused them did so before knowledge of the facts in question became public knowledge.

This looks obvious. It is not: not when a claim for scientific character was made, as long as this claim was understood as a claim for the possession of a genuine proof. My paradigm case is that of Luigi Galvani after whose name we call the process of galvanism that is the coating of metallic objects with a thin layer of another metal. He was one of the greatest contributors to science, yet he lost this status when he lost a scientific debate with Alessandro Volta. He resigned his academic post and left for pilgrimage to the Holy Land. When he returned to Italy he lived on the estate of his brother, never to return to public affairs. The fate of Sir Humphry Davy was worse. He made a mistaken suggestion to the British Navy and this made him leave England never to return. After 1905 things were different. Many scientists still change their views surreptitiously, but the admission of error is no longer punished, unless it is judged negligent and thus avoidable. Indeed, already in ancient books of laws—in both the Bible[2] and in ancient Greek and Roman books of laws, killing was considered a transgression when avoidable, not otherwise. Science, prior to 1905, was deemed the domain of avoidable error. Error in scientific texts was thus generally deemed unpardonable. It is still considered this way by many historians of science. This is embarrassing, and so in many histories of science errors of scientists are often still concealed. Thus, to take the paradigm case, the doctrine of phlogiston was deemed a prejudice. Lavoisier's alternative to it was then deemed demonstrated, although he said oxygen is the sole oxidizer. (He concluded from this that the oxidizer chlorine is a compound of oxygen and another element, one that he called murium. Later Davy argued that it does not exist.)

The idea of the siblinghood of humanity appears regularly in many cultures, perhaps in all of them. Yet as an idea systematically applied in diverse discourses it is unique to the cultures that include science as a central one of their ingredients. Richard Popkin notes that the biblical story of Adam and Eve signifies as it contains the theory of the siblinghood of humanity. He adds that in opposition to that theory, those who advocated slavery offered the theory of the pre-Adamites: the slaves are not descendants of Adam and Eve.[3] Significantly, this episode in the history of ideas has vanished with the toleration of slavery that the civil war in the United States in the mid nineteenth century brought about. It is an uncomfortable fact that until then slavery was common and deemed admissible as a matter of course despite its opposition to the thesis of the siblinghood of humanity. The reason is that this thesis is not easy to absorb into the general culture. It was the rise of high cultures like those of the ancient Babylonian and the ancient Egyptian that made it hard for the ancient Greeks (especially Thales, the first of the pre-Socratic sages) to view their own culture the highest. This opened the road to the development of the idea of the siblinghood of humanity. What characterizes humans is their relatively high intelligence. An example of it is the human ability to discuss the question, what is the best way to use our intelligence?

Michael Polanyi, one of the most important philosophers of the twentieth century, interrupted this discourse by an observation: many questions trouble intelligent, interested young people; they are not qualified to answer them or even to discuss them. In particular, they do not know which way to discuss significant questions is right or rational, much less why. Naturally, they cannot find the right tests that might enlighten them. So, they naturally consult some people in their vicinity. Of these people, some cannot tell them what rationality is, and some think they can, but in truth no one does. Yet what these young people want badly is help, not the right answer to an abstract question. They take it for granted that only those can help them who can answer the question at hand (what is rationality?) and to answer it intelligently.[4] This is a mistake. People who can do things—behave rationally or ride a bicycle or paint a picture—usually cannot describe their ability, much less explain how they came to acquire it. Their knowledge is personal: they know how to act. It is not verbal: they cannot articulate it. Some try, some do not, but none can. Efforts to articulate only distract one from the task at hand, which is to help the young people who do not know how to decide. Of course, only some of the people who have decided rightly and became masters can help them, but none can answer them. In other words, you start developing by listening. Is that listening right? Do you listen to the right master? Probably not. If you are lucky, you go to a good master and the good master may be able to help you. This is why all conceivable sorts of success—in particular, both artistic and scientific—come in workshops that able masters run.

Polanyi answers the question of rationality by observing the way people handle it. In practice, we all take chances. Yet we want to improve on practice, and for that we use our reason: we develop scientific theories; these improve our chances and create new options, some of which we gladly embrace. After all, this is how the modern world developed: scientifically. We have been there before: many thinkers have rejected this idea, some religious, others irrationalist, some of the latter Nazis. They criticize (scientific) rationality, namely, scientific method. A leading irrationalist of the twentieth century, Paul Feyerabend, denied the very idea of method![5] We are in a loop, because we are questioning the rationality of our discussion of rationality; that is why.

Fortunately, there is a way to break the loop. It is no doubt a great relief. It is this: first and foremost, try to be happy! How? Say, join a church. Which church? It seems we are back in a loop, but no: the answer is, join any church that you like. You can do some window-shopping. However, be careful: window-shopping can be endless! We seem to be back in the loop: when should we decide? There is an escape here: as long as you enjoy the window-shopping, continue; when you are tired of the window-shopping or are unable to continue it, just join a church.

This idea is often practically successful. Indeed, many of the young people who bump into the problem of rationality end the torment by joining a church; the church that most people join is the one into which they were born. So, we do know: the idea works. It has a name: *fideism*. (The word relates to "faith.") Fideism is a popular theory of rationality: you cannot solve it with no idea to rely on. So, choose any idea—arbitrarily—and then take it rationally. In brief, make a minimum concession to irrationalism and then be totally rational.

It is hard to see what the disagreement between rationalists and irrationalist can be. No irrationalist will recommend stopping all deliberation, and no rationalist will deny that at times stopping deliberation (and starting action) is vital. The dispute is philosophical. Criticism of tradition is ubiquitous. Some critics of tradition are radical; they recommend disregarding the authority of tradition and yielding only to reason instead. As long as radicalism was applied to science only (Bacon, Descartes), social institutions continued to function normally. The extension of radicalism (Diderot, d'Alembert, Mirabaud) to all human affairs heralded the French Revolution. The Reaction to that revolution (Burke, Hegel) demanded restoration of the authority of tradition, namely, to limit the application of reason by the application of caution. After World War II both radicalism and the reaction lost their edge with the rise of the innovative non-justificationist version of rationalism (Popper, Bartley) or anti-foundationalism. There still is a dispute about the relative weight of the two competing guidelines, reason and tradition. The locus of the dispute happens to be the quarrel between evolutionists and creationists in the education system of the United States that practices separation of state and church. Yet it is a poor example. For, this dispute concerns indoctrination (of schoolchildren); it has thus nothing to do with theories of rationality.

The dispute about the relative weight of science and tradition in everyday life is significant, but it is no longer philosophical (Wettersten). This is a clear case where philosophy proper has ceased to serve as an integral, significant part of the philosophy of life that it was once. If we want it to regain its position as a guide for life, then we should try to improve it. To this end, the starting point is the wording of the new non-justificationist rationalism or anti-foundationalism. Now, whereas non-justificationism is the name for a relatively clear attitude (namely, the appreciation of criticism and the readiness to change opinions and customs and anything else under its impact) under the banner of anti-foundationalism appear diverse attitudes, some serious some not. What is common to all participants in this discussion is the reserved appreciation of Bacon's demand to suspend judgment: we cannot dismiss all of our judgments and begin afresh. To take a conspicuous example, I cannot doubt for a moment that I am a human born to a woman, short of days and long of trouble.[6] Interestingly, when Descartes decided to enter his famous, world-shaking experiment of suspension of all judgment, he undertook to go on a pilgrimage to a famous shrine upon its conclusion. In reaction to the failure of the effort to suspend judgment, philosophers, especially William James, recommended the suspension of disbelief. This is odd. It is obvious that we usually do both when we look at a picture or at a play and when we read a story: if the experience is delusive, then we are psychologically imbalanced, yet if it is not, then we do not partake in the aesthetic experience—for want of empathy. For the experience to occur we must experience both; we do so in some back-and-forth moves. The same goes for the encounter with any new idea—be it scientific or metaphysical or any other—since the mere readiness to listen to it is already a measure of credulity; the readiness to invest the efforts necessary for any test of a scientific theory is much more so (Popper, *The Logic of Scientific Discovery*, §29).

It is thus obvious that even the classical theory of rationality is not clear, and from the very start. Yet historically it was the modern study of the very idea of proof that led to a great upheaval. Hume stressed that he used the philosophy of Locke to show that inductive proof is impossible. Kant claimed that the traditional concept of logic is too narrow, and he tried to develop an extension of it that he called transcendental logic (logic that transcends the old limits of logic). He failed miserably, but he also did something wonderful as he tried to limit the old logic as much as he could. He limited the form of sentences that the old logic could prove to two: "a = a" and "ab = a." These are tautologies (= repetitive sentences). Boole then developed his calculus of classes that he showed includes Aristotelian logic (corrected). This led Frege and Russell to develop formal logic proper. They did this in order to develop a theory of proof. This led to the view of proof as deduction from axioms—akin to the idea that is at the base of Euclidean geometry (Kurt Gödel). The general result from the new logic is one that Russell worded first, and Jaakko Hintikka repeated two generations later: *all and only tautologies are provable*. This is the newest version of Hume's disproof of induction and with it the disproof of the traditional theory of rationality, namely, that all and only the admissions of scientific theories are rational.

Philosophers of science still take it for granted that only the acceptance of scientific theories is rational. Rather than worry about their own belief in many theories that are not scientific, or at least to ask which ones of all these theories are scientific, they take it for granted that to begin with we are convinced that the theories we admit are scientific and that only then it is incumbent to try to justify this conviction of ours. This they view as efforts to refute Hume. (Since Popper and his followers consider Hume's criticism valid, the other philosophers of science simply ignore their output as irrelevant.) In other words, most philosophers of science still endorse Bacon's demand that researchers endorse all and only scientifically proven theories, and they still seek a theory of scientific proof that does not allow for Hume's disproof. This renders the theory of science a theory of proper or rational belief. Science then is a religion surrogate. This then is the radical theory of rationality that is the received view all vociferous declarations to the contrary notwithstanding.

NOTES

1. Bacon, *Novum Organum*, Bk. II, Aph. 19, 2012 (1620).
2. *Deuteronomy* 19:10–12.
3. Richard H. Popkin, "The Pre-Adamite Theory in the Renaissance," in Edward P. Mahoney, ed., *Philosophy and Humanism: Renaissance Essays in Honor of Paul Oskar Kristeller*, 1976, 50–69.
4. Michael Polanyi, *Science, Faith and Society*, 1964.
5. Paul K. Feyerabend, *Knowledge Without Foundations*, 1961.
6. *Job* 14:1.

Chapter 8

The Flotsam and Jetsam of Philosophy

The present study centers on the philosophy of life or practical philosophy. It has many known versions. Many philosophies claim to be practical. Are they? Which of them is most practical? By what yardstick should we compare them? By the yardstick of success. Yet we disagree about what success counts; there is financial success, political success, scientific success, the achievement of marital bliss or peace of mind, etc. We do not know which counts more. The default choice is, whatever the individual in question deems success should count as the measure of that individual's success. This looks unproblematic, but it turns out to be very complex, as scores of stories come to illustrate, from folk tales and parables to some of the most famous novels around. One may devote one's life to one cause and then change one's mind and try to move to devotion to another. In some cases, the shift is happy, at other times it comes too late.

Varied as the literature on this matter is, however, one item stands out as rather obvious: it is best to ignore philosophical disquisitions, as they lead nowhere and only complicate matters needlessly. They are thus at best a waste of time, and not a waste enjoyable at that: they are impediments. This is a very famous view. It is known as anti-philosophy or positivism.

One way or another, this looks like a serious obstacle. There is a way around it. Every devastating argument invite the search for means to circumvent it. There are too many options of circumventing every devastating argument. The practical question often is, which way to circumvent a given devastating argument is best? How good are any of the ways to circumvent a devastating argument? There are often too many of them to check and the way around some of them is a very wide detour. Before undertaking such a venture, one may want to make sure—to some extent, of course, namely, within reason—that it deserves the effort. Much learning is sheer waste. Think of studies of magic—for and against—for example. These fill libraries. Do they deserve study? We do not know. There is much magic that is obviously arbitrary. Think of the literature on reliquiae, on shrines that allegedly hold relics of the holy cross or of some other holy objects; think of the literature on holy graves that are centers for pilgrimage. If you wish to study these then perhaps you are a fool. Perhaps not: social anthropologists study magic with no interest in anything supernatural. Some of these studies of magic are very interesting. Magical practices are still prevalent even in the most enlightened societies. This information is an eye-opener, and it is quite relevant here.

Feyerabend suggested to his Afro-American students in Berkeley that they return to the practice of magic. It was a silly proposal, made casually just in order to elicit some response that might possibly be interesting. For, obviously, the suggestion made no sense since these students were no more magic oriented than were other American students. Admittedly, their tradition was saturated with magic; but this is true of all traditions we know of. Nevertheless, Feyerabend casual suggestion touched a nerve. It became the talk of the town in Berkeley. This led Feyerabend to the slogan he became famous for: anything goes. This slogan is not new: already Descartes noted that no idea is so objectionable that it has no advocate. Feyerabend went further and said; hence no idea is utterly objectionable. And he dismissed practically all objections to magic by observing that very few Western scholars are familiar with the literature on magic, which is, indeed, colossal. Moreover, innumerable items are there, awaiting in vain the attention of indiscriminate scholars. Every scholar can lose all sense of orientation by simply entering any university library. The wish to have some idea about its contents already overwhelms. More relevant perhaps is that Feyerabend was an Austrian and Austria a soaked in a magically minded version of Roman Catholicism, relics, miracles and eternal flames are everywhere.

There is no way around it: we need an agenda to serve as a filter that is essential for any act of choice. Polanyi says, we take some filter on faith, you remember, and we must do it, out of sheer ignorance. This is true. Yet his intent in saying that was to discourage independent thinking, which is objectionable. Our ignorance is vast, but our hope is that we can improve our thinking along the way, including our choice of filters. One rule of choice is useful here: keep the agenda open to change and examine it now and then. As we alter our agenda, we lose interest in some items. The problems regarding them may lose their value: we may want to investigate this matter. Of course, such investigations serve as time savers yet they also consume time. We prefer not to waste mote time on a time-saver than it can save, of course.

How mistaken Feyerabend was can be seen from the most obvious fact that one of the most important ways to improve the value of research is to cease studying obsolete problems. Perhaps I should not say that this is obvious, since so many researchers do just this, including most of the scholastics. But this happens even in the modern physical sciences, no matter how rarely. An amusing example is the study of atomic weights. The discovery of isotopes rendered obsolete all past measurements of atomic weights as now it is necessary to measure separately each isotope of every element that has them. Theodore William Richards, the first American to receive the Nobel Prize in Chemistry, earned the award in recognition of his exact determinations of the atomic weights of a large number of the chemical elements. Although obsolete, these still had some significance, he observed in his Nobel Prize acceptance speech, as the measure of the average proportion between different isotopes of a given element on earth. (This renders the results much less accurate than it looks, since possibly the ratio of isotopes of a given element differs in different layers of the crust of the earth. Clearly, he was on the defensive, and clearly the weakness was not only his: the Nobel committee offers prizes first and foremost to factual discoveries, rather than to theoretical ones. On this they entered into a conflict with Einstein as he was reluctant to accept the prize.) More striking is the waste of energy in philosophical study of obsolete problems. Were Feyerabend right, were his slogan "anything goes" valid, then the decision to cease studying obsolete problems would be no better than decision to continue studying them. This means that scientific progress is no better than scientific stagnation. Indeed, he said this too, justifying it by reference to the possible misuse of science. He even proposed that there should be a separation of state and science, akin to the separation between state and religion. Now the existing separation is not between ideas: this makes no sense. nor is it the separation between studies; this would be obscurantist.

Rather, the separation is between authorities, between political and religious authorities. This makes sense since religious authorities have to apply religious discrimination that political authorities are required to avoid. What are the scientific authorities? What do they discriminate against? The answer is, superstition: scientific authorities (are supposed to) discriminate against superstition. The separation that Feyerabend has recommended, then, would prevent of state from, say, proscribing the sale of prescription drugs in alternative medicine shops. This proscription is usually against the sale of poisons in the guise of medications, which Feyerabend would scarcely oppose (though his slogan, anything goes should make him oppose it) and the sale of narcotics, which he should object to (except for the case of children that is more problematic than the case of adults) as should we all.

Let me take an unusual example from philosophy proper. A central problem in philosophy was the problem of realism so-called. It is simple: is daily experience real or is it a dream of some sort? This is often called Descartes's problem, although the slogan "life is but a dream" is pre-philosophical and a part of Japanese lore, for example. Descartes famously developed his philosophy from the very beginning, with the intention to allow every stage to pass only after it has been proved. He therefore began with doubting everything and proving to his own satisfaction that he exists and getting stuck by his inability to prove the reality of what philosophers call the external world (namely, the whole world minus myself or minus all people's souls, as you like). It should be noted that his proof rested on the by-now defunct Aristotelian logic: he took his doubt as certain and by logic this predicate required a subject, the doubting ego. Already in the eighteenth century this led to doubting Aristotelian logic: Lichtenberg said, "it rains" is a predicate that does not require a subject. This is interesting. When Frege developed logic, he allowed predicates with any number of subjects, yet he ignored the zero number of subjects, as this renders every such predicate a statement. This is possible in ordinary discourse (like the Latin "vale!" or the English "Morning!" that is a common abbreviation for "Good morning!"), but it is of no use in mathematics, which was the major reason for Frege's effort to develop an artificial language of logic.

The inability to prove the reality of what philosophers call the external world was taken as a challenge; once, philosophers such as Denis Diderot deemed this challenge the most important task for any ambitious philosopher. Now most philosophers still struggle with this problem even though Descartes's program of creating science (namely, certitude) is by now rightly deemed defunct, to repeat, due to the contributions of Einstein and of Russell.

Worse than getting stuck with an outdated problem is, of course, the refusal to acknowledge that a refuted answer to it is refuted. The pretense that a refuted theory is unrefuted f is often justified by the fact that by and large technology employs refuted theories. The outcome of such conduct is dogmatism. It has disastrous results both practically and theoretically. Practically it is the inability to learn from mistakes, which means vulnerability to the damage due to their repetition. Theoretically, it is stagnation that is the boredom of having no new ideas to deliberate upon. The hazard of dogmatism also conflicts with Feyerabend' s slogan, anything goes. He was aware of it, and so he declared that no theory is ever obsolete. For a paradigm case he discussed Aristotle's philosophy, since it is a notorious fact that the Renaissance of research rested on the rejection of scholasticism, the philosophical school that took for granted that Aristotle's philosophy is impeccable. In Galileo's tremendous *Dialogue Concerning the Two Chief World Systems* (1632) the Aristotelian interlocutor finds it amazing that the speaker for Galileo has convinced him that Aristotle's theory of gravity is inconsistent. He says, if I cannot trust Aristotle, whom Dante called the prince of philosophers, then whom can I trust? Trust yourself, is the reply, the answer that Galileo offers to his reader. Does Feyerabend agree? Does his slogan help in any way increase self-reliance, or does it diminish it or take it away? It seems that the slogan of Feyerabend is but a pretext for irresponsibility, for the idea that there is no need to be responsible for your assertions since nothing you say is harmful. Yet Feyerabend has offered as an argument for his dismissal of science the observation that even science can be misused and that it often is. The insinuation is that science does more harm than good. Feyerabend told me publicly that as he gained hardly anything from frequenting certified physicians, he went to quacks instead. Privately he added that he only talked with them but never allowed them to treat him and he never took their advice. His famous slogan is, anything goes. If so, then his suggestion that science-based medicine is no better than quack medicine goes too. But then, also, exposing his assertions as disingenuous goes too.

A famous theory that many philosophers of science still cling to is the perception theory of Bacon and/or Locke. The great discovery of the time was the theory-ladenness of observations: there is no theory-free observation. It is unavoidable, said Galileo. Otherwise, he explained, as you walk in a moonlit street you will see the moon jumping from rooftop to rooftop like a cat. Galileo saw no problem here. Following Giordano Bruno, he acknowledged the value of observation but denied that they constitute proofs. As proof of the validity of a theory observation should not depend on theory, since in that case the proof is circular: theory and observation are interdependent. Most philosophers of science cling to the idea that science gains its validity from experience, since experience provides sense data, namely, information given to the senses that are merely passive when receiving them. (The word "datum" means given. It was Bacon's idea that we see correctly only when we are utterly passive, since Mother Nature does not deceive. These days all perception theorists agree with Kant: perception is active, hardly ever passive. Nevertheless, countless philosophers still cling to the idea that observation validates scientific theories. They do so because they take for granted that rationality demands that we justify our views and that only observations can do that.)

Locke is the thinker who reintroduced the idea of sense data to modern thinking. He wanted to square the Copernican hypothesis with empiricism, with the idea that scientific proof rests on observation. Bacon knew about sense illusions. He promised to offer a way to eliminate them from science.

He did not. No one did: sense illusions are unavoidable. Hence, all classical theories of sensations are by now empirically refuted, as is the theory that science rests on observations. In many universities, the departments of psychology maintain perception laboratories that house instruments that generate fascinating sense illusions of diverse sorts. Nevertheless, many philosophers still cling to Locke's theory of perception. And to his assertion that pure sensations are free of all error, that sense illusions come only from misreading.

Locke is famous as the philosopher who paid most attention to common-sense. And, indeed, some of his assertions rest on nothing but common experience. (For example, he said, just as courts do, we should admit a copy of a document as we admit the original, but not a copy of a copy.) Yet common-sense has no trouble with sense illusions: we recognize differences between a mirage and a normal perception; we have little trouble avoiding considering sense illusions true. Hence, the mistaken perception of the sunrise and the sunset is no sense illusion. This disproves Locke's theory of perception. Taking it as true, we must conclude that our distinction between mirage and true perception is a theory; and then we want to know if that theory is true. Philosophers of science cling to Locke's theory since they want to be realists: they want a science that describes the real world, not any fantasy. And they want this to be demonstrable. Data are given; only as given they are not question-begging and so they are demonstrative. They supposedly secure scientific realism this way. Yet the do not. They do the very opposite, as already Locke has noted but refused to admit.

Kant did not criticize Locke's theory of perception; his followers did. He said, we receive sense data even when we are passive, but science is active. He therefore denied that data belong to science: we render them scientific in two steps. First, we state them, we put them into words. Second, we generalize them. Data are passive, subjective; observation statements are active, objective—or rather inter-subjective. For, Kant gave up all claim for objectivity: he replaced it with inter-subjectivity: science is whatever we share in the intellectual world. That should keep science universal. Hence, Kant rejected every idea that he deemed particular, even Christianity! This way he was more Baconian than Bacon: he took scientism for granted: we can believe in nothing but in what science tells us. This is so alluring that people fell for it despite the appalling corollary from it that every one of us is born with Euclid and Newton imprinted in their brains! There was one catch: Kant could not assure us that the whole world is real. Solomon Maimon said, Kant's philosophy as a whole is one big hypothesis.

Under the coarse influence of Wittgenstein, some philosophers have claimed that this sentence, "the world is unreal" is meaningless, even though he himself confusedly argued both that it is meaningless and that it is false. The popularity of this idea in traditional Japan is interesting. It is a part of many mystic doctrines. These doctrines clam that people who follow their percepts are enlightened and enlightenment is the ability to perceive things on a level that most people do not achieve. Consider the way cats and dogs sense the world, as higher apes do, as ordinary children do, as ordinary people do in their sleep, in their reveries, in their waking hours and when endowed with a moment of blessed vision. These are levels of perception, each of which describes a level of reality. Mystics usually do not oppose science but see it as intermediate between commonsense and enlightenment. We need not discuss all this here. What practical philosophy invites us to do is to ask, is it worthwhile to leave everything behind and join a sect that promises enlightenment? Rationalist philosophers unanimously take it for granted that joining a sect—any sect—is a waste of time if not also positively harmful. They scarcely pay attention to it. Their attitude is dogmatic. The right attitude is obviously to discuss any interesting doctrine, be it of a sect or not, in the abstract, overlooking its adoption by a sect.

Kant insisted that although we have no access to the real—to the thing in itself, to use his idiosyncratic terminology—we must admit its realty. He took it for granted, hardly finding it necessary to explain why. Yet this issue, what is real, was major in traditional philosophy. Of course, scientific theories refer to the stars in the sky as well as to sticks and stones and cats and dogs. Yet a venerated tradition denied this. Ever since astronomers refused to admit the refutation of ancient astronomy, they developed an "as if" science. This looked fairly innocuous until Galileo declared, in 1632, that "as if" science is a form of cheating that destroys real science. The avowed reactionary Duhem, one of the greatest philosophers of science, developed the idea that all science is "as if" and he declared naïve the opposite idea of Galileo, the idea that science attempts to describe the real. He labelled the dispute as that between realism and instrumentalism. The instrumentalist finds it easy to dismiss the thing in itself of Kant! All this may be very interesting history, but it is sheer ballast all the same, instrumentalists should ignore it. Except that when instrumentalists ignore all practically irrelevant items, they are in error even from their own practical point of view, since we do not know what idea is practical, and since those curious about the world regardless of practical affairs happen to have made possible the most practical innovations. The paradigm case is Faraday's discovery of the motor and the dynamo that rendered the current use of electricity possible, although only after the added inventions of other pioneers such as Edison.

We are here squeezed between the idea that idealist philosophy should be dismissed as subjectivist and the idea that we should not ignore ideas just because they seem useless since they may turn out to be interesting and even useful. The way to overcome this discomfort is rather simple: we should be aware of idealism and of our view of it as useless since it is subjectivist. We can leave it at that, to allow anyone interested to take it up. I would go further and say, here is the great error of the enlightenment movement: still under the influence of obscurantist, intolerant elements in some traditional religion, its (Baconian) custom was to dismiss erroneous ideas as misleading. Not only the fear of being misled is exaggerated; forgetting past errors amounts to forgetting the past altogether. And indeed, the enlightenment movement is known for its lack of interest in history. It was the romantic reactionary movement that returned to the study of (political) history as a tool for enhancing its conservative attitude. But critical historical studies need not be conservative; they can hardly be that. The fallibilist attitude to history is the view of it as the store of all past errors, wise and foolish. Philosophers of science still speak of the rational way to entertain and discard ideas, which is subjectivist, and rooted in the alleged need of the rationalist to justify. The very attitude is subjectivist. The fallibilist, objectivist attitude is different.

For the fallibilist to entertain an idea is to examine it critically; discarding it is to store it in our social memory so as to be able to discuss it critically on some future occasion. The rejection of an idea is subjectively to consider it false; objectively it is always a true-or-false potential object for examination. Thus, a cardinal part of fallibilist philosophy is the idea that no refutation is final, that it is always given to reassessment. The once critical rationalists Feyerabend and Imre Lakatos[1] found this a weakness in critical rationalism. This shows that their rationalism, while they advocated it, was not sufficiently critical. This is not surprising. Criticism was always a part of all rationalist philosophy since limiting it amounts to dogmatism; yet it was always taken as the appetizer. The great nineteenth-century critical philosopher and historian of science Whewell was very clear about that. He said, researchers invent hypotheses, and these are likely to be false. They should be refuted; only the idea that survives refutation is scientific (i.e., proven). He gave as an example Kepler's researches, taking for granted that Kepler's famous three laws are true. When he then discussed the chauvinist and obscurantist idea of Hegel that Newton had plagiarized his theory from Kepler, he noted that by Newton's theory Kepler's laws are true only in a two-body system, not in the solar system, he was on the verge of discovering a critical rationalist idea, but he balked. All this remained vague as long as the resistance of Newton's theory of gravity to all criticism was taken to be proof of its truth. When Einstein was willing to give up certitude, he offered a radically new view of the past: in physics the old theory is a special case and an approximation to the newer and better theory. This idea is at times applicable outside physics, but always less obviously so.

We should remember our past mistakes: they comprise our heritage.

NOTE

1. Imre Lakatos, *Proofs and Refutations*, 1976.

Chapter 9

The Romantic Idea of Reason

Some people say, the use of reason is better avoided: one should follow one's heart rather than one's brains. This idea is as odd as the idea that we should not use our hands. Nevertheless, it makes some sense: we often say, your hesitance makes good sense, but life must go on, and so you should not put much weight on considerations of difficulties that any venture faces. Similarly, we often say, do not dismantle a watch as you may be unable to put it together again; do not introduce this or that social reform as it may destabilize society without the possibility of turning the clock back. The point of all such suggestions is that as both the brain and the hands can be misused, we should hesitate when introducing social change. Some conclude from this that all social and political reform should be banned; others conclude from this that reform should be introduced only when the need for it is too great to ignore; still others conclude from it that every proposal for any social or political reform should be discussed and put to a vote. This is the practice of the modern liberal democratic parliaments. In some of them parliaments have two houses, in others supreme courts have to ratify new constitutional laws. Radical politicians often declare such procedures too slow for comfort. All this shows how many ways there are for impeding the rule of reason. Advocates of the unlimited use of reason are familiar with these limitations that we put on it for fear of doing harm, and they suggest that there are ways of avoiding error altogether. This idea, the idea that there is a way to avoid error, is called scientific method. The idea that scientific method is possible, then, conflicts with fallibilism. This fact causes much confusion. The idea that there is a scientific method is extreme. Its opposite is the idea that there are no rules, that, as Paul Feyerabend has put it, anything goes.

Nature is rich, we cannot put Her in chains (to use Bacon's apt metaphor). Still, although scientific method in its Baconian sense (of error-avoidance) is impossible, there are many other methods, in science and more generally in civil life. The objection to all rules that is Feyerabend' s hallmark is thus anarchist: anything goes. As Popper has noted, one can easily read this slogan as the permission to advocate Fascism if not its advocacy tout court. He even suggests that much of the texts of Feyerabend go this way. For, Feyerabend did say that even the experience of violence is enlightening so that its advocacy is not devoid of merit! This is terrible. Does that mean that even the use of reason must be limited? Yes, of course. Such is the moral of the 1896 H. G. Wells story "The Island of Doctor Moreau"[1] and the rules known as The Helsinki Rules that require of all medical and social science researchers to follow: those who use humans for experiments must make sure that they do not violate the freedom and dignity of their subjects. This is a clear limitation on the use of reason in research, one that was admitted only early in the twentieth century.

The enthusiastic advocacy of the use of reason by the Enlightenment Movement led to the disregard for traditions altogether and thus to the disruptions of continuity of the development of the commonwealth of learning that the French Revolution still symbolizes.

Now the French Revolution was a failure: it ended with the French Terror and the Napoleonic wars. Yet it remains a beacon of enlightenment and thus it expresses an ongoing concern with its ideals: liberty, equality and fraternity. By now, incidentally, the ideal of fraternity is almost universally dropped. (Martin Luther King repeatedly said, love thy neighbor. Stand-up comic Dick Gregory objected: he wanted his neighbor to respect him; love is optional. This chimes with Popper's restriction: he said; "help your enemies; assist those in distress, even if they hate you; but love only your friends."[2] There remain the ideas if liberty and equality. How they are applicable to political reality is hard to judge. This is no cause for despair, however: it invites political experimentation.)

The question, "What is the place of tradition in reason?" remains open. The honor that Descartes receives as the first modern philosopher he won because he asked it first, and also because he answered it in the fashion that he learned from Bacon: forget the past: it is a bundle of prejudices. Most current critics of Descartes center their criticism on his starting point: it is impossible to forget the past; at the very least we inherit our language from our society. Moreover, said Sapir and Whorf, the particular language one speaks influences the way one thinks about reality. Some commentators take that to mean we are locked in the cage of our language and cannot escape it. If anyone held this hypothesis, it was Wittgenstein in his second phase. In any case, this is obviously a false hypothesis: we cannot free ourselves of all the prejudices that our mother language or our acquired language imply, but we can improve matters piecemeal. An idea along these lines is articulated in Popper's 1949 "Towards a Rational Theory of Tradition."

Piecemeal chipping away at the cage of language and other prejudices leads us to another question: are people naturally critically minded for this to happen? Possibly loyalty is natural, and possibly it becomes dogmatism under the influence of the (erroneous) received idea that loyalty involves staunch adherence to some given beliefs. That this idea is received within the rationalist camp is a bit hard to swallow. Yet unfortunately it is so. Most philosophers of science still ask, what is the degree of credence that a theory deserves in the light of all the information available? The answer is, of course, all and only scientific theories deserve utter credence, and other ideas one should dismiss as superstitious. That answer, however, raises the question, What theory is scientific/deserves credence? This is the problem of demarcation of science that Popper's revolutionary theory of science came to solve. It is also, however, a source of great confusion. For, whatever is the answer to the question, "a theory is scientific if and only if it is x," current attitudes to science translate it to "a theory is credible if and only if it is x," whereas Popper's theory ignores credibility altogether. The question Popper raises is, what does the acceptance of a theory means if it is not accepted as an object of belief. Popper answered it repeatedly his *Logik der Forschung* (1935)/*The Logic of Scientific Discovery* (1959): in science, the acceptance of a theory is its acceptance as an object for empirical tests. Its rejection then is likewise not its rejection as an object of belief, much less the decision to ignore or forget it but the decision to view it as already refuted and thus as no longer an object of test. That decision, of course, is only a default option, always open to reversal. Popper's proposal is to consider a theory scientific if and only if it is open to empirical criticism, where the openness is a matter of principle, not of the decision to test it: here there is an order of priority, since it is impossible to test at once all extant testable theories. Moreover, the proposal is to consider a theory refuted as the default option when an empirical report is made of an observation that contradicts it, and not to amend it *ad hoc* in order to save it from refutation. (The amendment is always easy to make: if a universal statement is refuted that is a corollary from a theory under test, then it is easy to replace its expression "all x" with "all x except y," such as in "all swans except the Australian ones are white" or "all swans except the Brazilian ones are white" or even "all swans are white except the Australian ones or the Brazilian ones."

Criticism is then always of the same ilk, subjective or objective: given the information available, the subjective version goes, a person may err and allot a given theory the wrong degree of credence. This is obviously irrelevant if not downright false: in common practice, it is very different. Usually, critics argue against claims that some theories or observations are true or against recommendation made under some circumstances. This leaves it for the targets of the criticism to decide what to believe in or how to act—assuming that these beliefs are under control. Some control over beliefs is possible, but only marginally so. The incredulity that a report meets with may be the subject of a discussion. But not always. One day you wake up and find that your faith is gone, said George Orwell, and you can do nothing about it.[3] Even the little that is given to decisions is given under very restricted conditions; in any case, such decisions require autonomy and some familiarity with the situation under discussion, yet autonomy is alas still uncommon. And so, no, most people are regrettably not sufficiently critically minded; we are too credulous.[4]

The Romantic movement declares it impossible to make common people autonomous. Only a few are blessed with autonomy; they are the geniuses. A genius may break the fetters of his society that is his cage. Now a person with no social fetters is alienated. (The word is a neologism that pioneering psychiatrist Philippe Pinel introduced as the politically correct reference to the mental patient.) The genius may return to the fold, however, when his outlandish ideas are absorbed there, and he thus becomes a hero. This is possible since truth is relative, but this relativity is given to a society—not given to the individual to manipulate (with the exception of the genius). In brief, the romantic movement is conservative with limitations: it allows for islands of the enlightenment movement here and there, and so social revolutions here and there, only when change is unavoidable. Social philosophy is conservative almost everywhere. Its rationality lies in the claim that tradition is the store of accumulated knowledge. The Enlightenment Movement was radical: it required giving up all tradition, which is impossible. This led to the reaction, which is the recommendation to return to tradition with a vengeance. It is important to notice that the reaction is to radical ideas: it is the awareness that radicalism is dangerous and invites legislation to emaciate it. Once reaction is an option, traditionalism naturally tends to disappear. What remains, then, between radicalism and the reaction is reformism. The commonest version of reformism is parliamentary democracy. Oddly, although parliamentary democracy is the paradigm of reformism, it often houses more extremists than compromisers. Compromise is more often the result of clashes between extremists than a philosophy. For, the philosophy that allows compromise as a principle is fallibilism. This is odd. The idea of fallibility is ancient: "Put not your trust in princes, nor in the son of man, in whom there is no help," says the Psalmist (146:3), trust the rock of the ages. Yet the Age of Reason promised infallibility. How? This is the problem of induction. It is insoluble. It is insoluble even when replacing proof with proof-surrogate.

The failure to solve the problem of induction was always the argument that the traditionalists used, not against reason but against rationalism. For traditionalists (other than the obscurantists, with whom one can hardly communicate), reason may serve tradition or destroy it; hence, in itself reason is neither good nor evil, which is true of all weapons. This obviously commonsense idea was adopted by the commonwealth of learning only after Hiroshima. Before that, the rationalist claim was much stronger: the use of reason cannot be harmful. Of course, even before Hiroshima misuses of reason were known since science always serves war. (Early in the scientific revolution Boyle purchased the newly acquired patent on the machine gun in order to prevent its use. The inventor sold the patent to him and then to the English and to the Dutch during the First Anglo-Dutch War of 1652–1654.) Yet the claim was not unreasonable that on the whole science has improved matters. Only in the twentieth century did it become hard to sustain this position. The same held more generally for the problem of evil. That evil is ubiquitous hardly needs saying, yet it was Auschwitz that tipped the balance and made it hard to swallow the excuse that on the whole science has contributed more evil than good.

The philosophy of the second half of the 20th century looks increasingly divided between Popper and Polanyi. Popper viewed science as an utterly free market though he refused to view the economy as even possibly utterly free. He viewed the free open market of ideas as essential for science and so as in need of political guarantees. How much freedom is essential? He did not discuss this question, though he knew of course that early modern science is older than democracy. Polanyi viewed science as an autonomous club, in principle, not quite in fact. He did not discuss the question how near the ideal facts must be in order to insure the very existence of science. He viewed its autonomy as freedom from outside constraints. Internally it has a leadership, he said, unelected yet accepted by a stringent consensus. He was in error. The scientific consensus is not binding and science is no autonomous club: it does interact with society at large, as it must. To that end it has not one but two systems of dual leaderships—social and intellectual. Gellner viewed the latter as the underground. As these are not identical, the powers of social leaders may diminish upon retirement and those of the intellectual ones may increase that way. The ideology of science plays a significant role, though no one knows what exactly it is: the traditional enlightenment philosophy was too idealistic. It went up in smoke in Auschwitz. With no public discussion, the new ideology is a modified version of the traditional, enlightenment one. It thus ignores the impact of the military and of the market on the commonwealth of learning. That was traditional and reasonable. It ceased to be so as rude intervention in its life came in forcefully with nuclear weapons and with other, somewhat less destructive science-based technologies.

Whenever thins get tough, romantic commentators harp on the memory of the good old days. Now it is obvious that the good old days were not utopia. Proof: the reference to the utopian good old days is to the golden age that is frankly mythical. Thus, when Cervantes narrates the story of the golden age, he puts it in the mouths of shepherds who speak of it as of an ancient myth. Even the story of the kingdom of King Solomon (beginning of *First Book of Kings*) is not taken to be sheer myth, So, as the good old days are not even a near utopia, do we know that they were better than our days? Better for whom? Given the tremendous inequality and the scarcity of improvement of social status, hankering after the past belongs to people who imagine that had they lived in the past hey would have possessed a high social status. In other words, romanticism, or hankering after the past, is often sheer illusion. This is not always the case; some people hanker after the past since they like a certain quality that they appreciate and miss, and that was more prevalent at a certain period that they envy, such as chivalry. Yet the medieval chivalry that they rightly appreciate came at a high price, such as dueling, that they tend to ignore.

Still, the right attitude to all tradition, to repeat, is the one outlined in Popper's classic paper on traditions: it is that of critical appreciation.

Some people advocate the avoidance of the use of reason. This is as odd as the idea that we should not use our hands. Nevertheless, it makes some sense: as tradition is the fund of our experiences, whenever you find tradition erroneous, do think again: you may be in error; you may have misread tradition and you may oppose it on the strength of erroneous information or irrelevance, namely, you may have taken things out of context. This last point is very dangerous. Since no two contexts are identical, every generalization is suspect! When Boyle discovered this, it disturbed him, since he found it difficult to declare, as he did, that in every conflict between theory and experience the default option is to let experience win. What context we deem irrelevant is theory-laden; hence, every observation is, as Galileo has argued. By the way, Irwin Schrödinger said, the sun looks no bigger than a cathedral; the assumption that this is true leads easily to the conclusion that the distance between east and west is less than a day's walk! Since theory and observation are interdependent, then all we can do to improve their reliability—or rather to reduce their unreliability—is to pitch them against each other, as was recommended already by Democritus of old.

This looks insufficient, and so the Romantics declare unavoidable the reliance on tradition. Since both Polanyi and Popper admit this, it is hard to distinguish between them. Clearly, Polanyi stresses that we cannot articulate all of our knowledge and Popper refers only to the articulated part of it. Yet they agree that no articulation of any knowledge is complete and both agree that criticism is vital for science. The difference is in the stress. Whereas Polanyi stresses the significance of instruction as going beyond what is explicitly learned from the instructor, Popper stresses the importance of the critical attitude toward even the best of what you have learned. If there is a romantic aspect to Polanyi's teaching that Popper rejects it is that he dissuades certain kinds of research—research that is too independent of tradition—as futile, Popper leaves open the details of the process of learning. In particular, he admits that learning includes a dogmatic element, and he leaves open the question, how weighty is the dogmatic part of the process. Polanyi, by contradistinction, stresses that tradition is weightier than criticism: when the latter gets too significant, the dangers of losing touch with tradition become imminent, and it is the danger of utter futility. To the extent that Polanyi discourages criticism, he seems to me a romantic. I have discussed this with him, and he admitted that possibly he was too much of a romantic.

What is all that to any normal scholar? All normal scholars can improve their situations by the acquisition of more knowledge about the questions that they care about, and even the acquisition of the knowledge publicly available is inexhaustible. To reiterate, knowledge appears in statements that are context-dependent, when one cannot possibly articulate all the contexts on which one's knowledge depends. Already the early nineteenth-century German philosopher Hegel noted that the interconnectedness in question makes any knowledge of any item an all-or-nothing proposition. From which he concluded that you know nothing and that he knew everything. As he put it, your ideas are subjective (you think that they are true but they are false) and his are objective (true).

So, what is all that to any normal scholar? Probably nothing. It is ballast and it is confusing. Is there any value to it? First, as a critique of the Enlightenment Movement it has historical value. This, however, is of no interest to the normal scholar. Second, the idea of interconnectedness is important. Is it specific to the Romantic Movement? Perhaps not, but the idea is specific to the Romantic Movement that hence we can never know whether it is true, that even if our view is successfully applicable to all the cases that concern us it may deviate from the truth however slightly. Now we can say, what is an unnoticeable deviation from the truth to any normal scholar? The answer that Einstein has provided is remarkable: taking care of the small deviation may lead us to domains where that deviation becomes large and then it may become significant in a manner that we could not conceive. Let me remind you that one such small deviation is the small instability of the atom that has occurred to Einstein, one that magnified became the nuclear weapon that did not occur to him even in his wildest nightmares.

Some people advocate the avoidance of the use of reason. This is as odd as the idea that we should not use our hands. Nevertheless, it makes some sense: in times of stress we often say, do not think, act. If in a moment of stress, you do not know how to act, you may have to act anyway. Situations do exist that are so intolerable that almost any action is better than none.

These situations are known as being with the back at the wall. Trapped in fire or in enemy territory one is definitely in the position that requires immediate thoughtless action. Under such conditions, deliberation is fatal. It is well known that politicians are prone to deliberate even under such conditions and bring about disasters. Clearly, in such conditions, when any public action is better than hesitation, whatever the public action taken is, it must be coordinated and so it must be led by an individual. At times, the individual just happens to be there. Lech Wałęsa was the individual who led Poland in a moment of stress merely by virtue of having taken a position at the helm. Such events are rare since it is not enough to take the helm; it is not enough even when making the reasonable decisions: one needs to command obedience. How this happens no one knows. Nor is the result of such an action always felicitous. The case of Wałęsa was, since the Soviet Union was on the verge of collapse and he was a reasonable individual. The ayatollah Ruhollah Khomeini was a disaster for his nation simply because he was the wrong person: he was a fanatic and yet the nation preferred him to the Shah, probably rightly. Thus, even though it was right to take a chance and follow a questionable leader in a time of stress, not thinking is dangerous. But this is anyway obvious to everyone trapped in fire or in enemy territory who bursts out with no knowledge of the terrain.

The romantic philosophy recognizes danger and at times even embraces it. "Live dangerously!" is a famous Fascist slogan. But the romantic philosophy also expects success. The last radio broadcast of the Führer, the one in which he finally admitted defeat, he accused the German nation for it: the defeat proved the nation not deserving the greatness that the Führer was going to bestow on them. This kind of thinking, finding an excuse that always and in principle renders one never guilty of irresponsibility or even of error, is endemic to a kind of political thinking known as historicism. History has bestowed on the German nation greatness; it has proven itself not yet worthy of it, and so it ran into defeat but the future still awaits success. This is how the leading German philosopher Heidegger could remain a follower of the Nazi ideology to the end of his life, racism and antisemitism, defeat and Auschwitz notwithstanding. There is a romantic specimen for you!

It is cheap to take such a poor example as Heidegger. So, let me mention Max Planck. He could not stoop so low as to join the Nazis. When they accessed power, he asked for an audience with the Führer and had his wish granted. He requested that the Jewish origins of his *protégé* Einstein be ignored so that he would be allowed to continue his membership in the Berlin research institute uninterrupted. The Führer threw a temper tantrum. The wish was denied. Planck retained his hope that the Führer would come to his senses. He never did, of course. Einstein wisely did not return to Germany from a trip. Planck asked him if he needed financial help. No, replied Einstein: I have saved money precisely for this eventuality of authorities wishing to destroy his people. This is probably the first (private) reference to the Holocaust ever: he knew what the evil of romanticism is. As to Planck, he stayed in Germany to the end. He never made any public reference to the Holocaust. Einstein, who always treated him as his patron, did not reopen communication with him. He addressed the German physicists generally instead. He said, he would not judge their conduct under the Nazi regime, but their not having apologized and offered compensation afterwards made him decide never to return to Germany.

The German philosophical community fared much worse than the German scientific community. The popularity of Heidegger speaks volumes. He was a member of the Nazi party and viewed its ideology as reactionary chauvinist romanticism. He was the first professor who entered the classroom in uniform and with the Nazi salute. After the war, when some public figures who were advocates of democracy were appointed by the Allied occupation forces to public offices, he stayed out of the system and kept advocating his romantic ideas. He had a tremendous influence in Germany, although his followers found it hard to say clearly what idea he contributed or even what he stood for. Foreigners expressed admiration for his philosophy. This was a compliment too flattering to ignore. Why is he admired despite his frank anti-semitism and Nazi sympathies? Many authors have attempted to answer this question in simple language. Perhaps this is simply impossible. Heidegger himself said it is not: he said to be a philosopher you must know the German language. By which he meant his scarcely translatable German Jargon that sane people should better shun—together with his romanticism. What the allure of romanticism is includes the proposal to transfer individual responsibility to the fearless leader that was called eleutherophobia (fear of freedom). Recent political history shows how dangerous this can be.

NOTES

1. H. G. Wells, *The Island of Doctor Moreau*, 2020 (1896).
2. Karl Popper, *The Open Society and Its Enemies*, 24, III, 1945.
3. George Orwell, *A Clergyman's Daughter*, 1935.
4. Samuel Butler, *Erewhon*, 1872.

PART III

The New Enlightenment

Chapter 10

Minimal Standards in Ethics

Every culture includes the awareness of its particular nature. It naturally recommends itself as superior to all other cultures. The starting idea and the backbone of Greek philosophy is the distinction between nature and convention that includes the recognition of the arbitrariness of my culture, including the claim that my culture, and it alone, is the right one. Already Abu Nasr Al-Farabi, the first to revive Greek culture in the Middle Ages, noted wryly that the uniqueness of Greek philosophy is its dismissal of every other local philosophy due to the recognition of its arbitrariness. Yet one of the great ideas that the Greeks made part of the Western heritage is the distinction between the descriptive and the prescriptive. This distinction is so much taken for granted within the scientific viewpoint and as a major part of it, that we tend to forget that it does not hold for the magical worldviews that are much more prevalent than the scientific one.

The descriptive aspect of Greek philosophy was the rejection of the local worldview which opened the road to the rise of science. The prescriptive aspect of it was much more problematic. It rests on the descriptive claim that the conflict between society and individual is irresolvable. Those who ascribed priority to the interest of society—the collectivists—were conservative; those who ascribed priority to the interest of the individual were radicals. Their disagreement was primarily but not only prescriptive. Descriptively, the radicals deemed social science be inherently individual psychology, whereas the collectivists took sociology to be the basic science. Already Plato found this intolerable and he dreamt of a society in which there is no conflict between individual and society. He seems to have wished it to be neither conservative nor radical, yet it turned out to be hyper-conservative, namely reactionary.[1] Thus, there are two prototypes of moral philosophy, the collectivist and the individualist. The former bases ethics on tradition as it is observed, this giving weight to the descriptive and the latter on the demand for individual autonomy, this giving weight to the prescriptive. Plato's effort to bridge the gap between them was emulated repeatedly in the utopian literature. As H.

G. Wells has noted in his discussions of the diverse possible utopian regimes, since the most significant demand of a utopia is that it resolves the conflict between individual and society, it becomes impossible to say of a successful one whether it is collectivist or individualist. Popper's response to this is very impressive: since utopia is heaven on earth, all attempts to create one must fail; they then tend to be hells on earth. This true statement we should attenuate: despite their failures, the utopians managed to improve the quality of ordinary life. True, this success is largely due to the rise of reformism, which comes to replace the radicalism, or the normal large-scale reform, that utopianism is. Nevertheless, reformism owes its very existence to utopian dreams, both biblical and philosophical, and this is not to be ignored.

Reformism, too, is both prescriptive and descriptive. The failure of utopianism that has led to reformism has also led to the idea that since the conflict between individual and society, as conflict in general, is never fully resolvable, all reduction is impossible—the reduction of individual to society as well as the reduction of society to individual. Indeed, the very disposition toward reductionism was due to the recognition that the assumption that both individual and society are basic implies the inability to eliminate all conflict. The odd thing is that long after the recognition prevails that the prescription of the elimination of all conflict was doomed to failure, the prescription for social science to recognize this was delayed. The first to notice this explicitly was Simmel. the leading sociologist Durkheim was frankly a collectivist, wrongly identifying the collectivist theory with the assertion that sociology is not reducible to psychology. This very identification was the target of Simmel's criticism. Attacking him personally in lieu of answering his criticism is the paradigm of the deterioration of science and its noble claim for impartiality.

The net gain from all this is the conclusion that although there are three kinds of descriptive social theory, the radical collectivist and the radical individualist as well as their reformist synthesis, in ethics there are only two prototypes: the collectivist that is traditionalist, namely demands of the individual to heed tradition (at least as the default obligatory option) and the individualism that demands moral autonomy. The third option, advocated by Russell, Einstein and Popper, is to heed tradition while attempting to criticize it and to improve it in the light of criticism. They even declared this attitude sheer common sense.

The prime distinction between nature and convention, then, has undergone a major transformation. The advocacy of nature, both descriptively and prescriptively, aims too high. We do not know what nature is, and any idea about it is conjectural and will remain so. The advocacy of convention, however, again both descriptively and prescriptively, aims too low: we should be critical of it and try to improve it. It thus ceases to be utterly arbitrary: we move between the utterly objective and the utterly subjective. Our image of reality is no longer black and white, but rather all shades of grey.

Thus, our moral theories still vary from society to society, but criticism and resultant improvements bring together the diverse moral views. This should not imply that one day morality will be utterly uniform: new moral problems appear and lead to new solutions and thus to possible new diversities that we may later attempt to iron out. This tallies with Popper's general idea of rational thinking: it leads to possible solutions to pressing problems; which solutions raise new problems.

All this concerns ethics in general, with no specific idea as to its contents. Are there any requirements of any ethical doctrine in particular? As there are two kinds of ethics, individualist and collectivist, each with its own general requirement (moral autonomy and taking tradition as the default option respectively), clearly, the individualist has a heavier task when coming to details. The individualist alternative to the collectivist general rule to heed tradition is the general rule to value the life of every human being as such. Kant, the fairly generally recognized conscience of individualism, took this as the only categorical imperative. His philosophy is open to challenge, however. He accepted as a matter of course the identification of rationality with certitude. He even went so far as to recommend a ban on the publication of any hypothesis! He took only categorical imperatives as given and contrasted them with conditional imperatives. His fame was largely due to his dogged consistency, yet he had no way to decide that what he deemed a categorical imperative is not conditional. It is well known that he declared the proscription on lying categorical. He even proved it in his own way: were lying generally practiced, no assertion will be believed and so there would be no lying. This is silly, since truthfulness is never fully practiced. This is answerable by the claim that the rule is not the mere proscription on lying but the demand that when it occurs it should be met with proper penalty, hence as long as lying is penalized, the rule is kept.

Moreover, a sense of guilt may be considered a penalty of sorts. And yet only when lying becomes widely practiced with impunity does Kant's argument hold, not when it is practiced sporadically. This black-or-white character of Kant's philosophy is what gave him the reputation of a careful systematic thinker, yet it also made him the object of ridicule. Russell has offered the strongest refutation of Kant with the tritest counterexample he could think of. Kant said it is forbidden to lie to murderers in effort to rescue their victims. Russell said, he lied to fox hunters and felt no guilt. More seriously, Kant raised a terrific problem.

Since paternalism—helping people without their asking for help—is forbidden, teaching is forbidden. How then he asked, can we instruct children against their will? (He evidently could not imagine schooling with no coercion. Not only is this the practice in free schools: it is part and parcel of Jewish tradition.) He answered: children are not quite human—since they cannot respect the categorical imperative, since they have no discipline. Hence instruction is chiefly, though indirectly, the instillment of discipline, to cease the moment the pupil has acquired it.

The minimum standard of ethics, then, is the ability to follow rules. This is indeed generally recognized: the demand to do the impossible is always and in principle invalid. The trouble is, we do not know what is and what is not possible, much less can we offer a criterion for this. The paradigm case here is the license to kill in self-defense. Whenever such a case turns up it is examined with reference to the details of the circumstances in which it has occurred with no reference to any general rules: juries have to decide whether there was another way to defend oneself than by killing the attacker. But this is only one kind of absence of general rules. Often new situations occur that raise new moral problems. We ty to forge new rules about them. The paradigm case here, incidentally, is organ transplants.

Nevertheless, the universal moral rule stands: do not require the impossible. This is too lax: there are options that we laud but refuse to demand. We call them supererogatory, namely, acts performed above and beyond those prescribed as obligatory. Moreover, thus far we have referred only to obligations that apply to individuals—autonomous or obedient. This has left out other moral duties, to professionals—professional ethics—and duties that collectives hold together. These are much more complicated: what is the duty of an individual who shares a duty with some collective? Perhaps the simplest yet the hardest duty is to leave a collective once it has crossed some red line.

We have here a few kinds of items to discuss: collective duties, duties of an individual to a collective—be they voluntary or those that come with the territory willy-nilly—and red lines.

The idea of the red line is the extreme opposite to the supererogatory. Both of them the literature sadly neglects. The same holds for the gray area. How much should one reveal to others during negotiations to avoid the charge of deception? Interestingly, at a certain stage the Friends, known for their high moral standards, decided that under normal circumstances, any standard haggling—negotiations as to the price of a regular transaction—is immoral since it is normal sellers know what is the minimum price at which they are willing to sell. And so, they decided collectively to fix their prices and to advertise them. It turned out that this is a wise policy and so it became the standard in normal modern markets. Needless to say, this is not always so, as the stock exchange makes all too obvious. This leaves a vast gray area. Whereas clearly the price of gilt-edged stock is public knowledge and the price of rare objects such as artworks, is made known only in auctions, many items and many commodities have variable prices, to be determined by auctions, which is then a kind of haggling. And then often sellers are unable to decide whether to fix a price, and if so which price, or to haggle, and then whether in one manner or in another.

There are here at least two kinds of problem that enable one to be more moral or less. The one is, how often one deviates from the norm, and the other how strict is the norm that one adopts—between the shady and the supererogatory. And we have said almost nothing as yet about the norms themselves that differ from one society to another. Consider, for example, the standards of proper sexual conduct that were accepted in our own society but a century ago and our view of them today. In particular, the attention to the dignity of women has radically altered—admittedly not enough, yet quite significantly. My own mother met my father first under the wedding canopy. Marriage then had to be consummated at once. We can hardly avoid today considering this custom legally established rape. Many feminists condemned the marriage institution as such, and in very strong terms. That this institution is under rational debate is obviously beneficial, and only obscurantists can oppose it.

This is not the place to discuss sexual morality. The principle under which we should conduct such a discussion has been set in the very opening of Russell's *Marriage and Morals* of 1929: there is no sexual morality except for morality applied to sexual conduct.

Hence, morality is universal rules of conduct applied to diverse situations. Of course, both the rules and their applications may be problematic, especially, we remember, when applied to utterly new situations such as organ transplants. The rule that we have thus far mentioned is that of respect to human beings as human beings. This may look obvious to some people, but this is very questionable. For an obvious example, we consider slavery immoral, yet it was legal even in the modern world until the American civil War in the middle of the nineteenth century: it wasn't until the middle of it that President Lincoln declared the freeing of slaves a war aim. Nor did slavery vanish after that war. Marx went further and deemed modern factory workers slaves. Reading Friedrich Engels's study of the conditions of workers in factories at the time makes it hard to disagree with Marx. Engels himself did not appreciate his study much, even though it was a pioneering sociological study: he published it in his native German and had its English translation published only half-a-century later. When the QWL (quality of work life) movement began there were objections to it on the grounds that workers are free to leave the work organizations that they dislike. This shows how much need there was to raise the quality of work life in order to protect the dignity of workers. The enemy of the workers in the modern world, it seems to me, is the work ethic so-called that takes the descriptive biblical text as prescriptive: "By the sweat of your brow you shall eat bread."[2] The result of the folly of endorsing this curse is that every time a robot gains a new portion of industry workers fear unemployment instead of looking forward to the reduction of the workload that this progress enables. The basic idea that wins almost universal consent here, to repeat, is the alleged irresolvable conflict between individual and society. Obviously, this is not the case: living within a flourishing society is in the interest of every individual. To respond to this deadly criticism a distinction is made between the individual's short-term and the long-term interests: in the short-term interest, one is lazy, whereas in the long term it is to work hard in order to secure a prosperous future. This reference to the future prosperity is called delayed gratification. In most cases it is a lie: often enough one dies before enjoying a lucrative pension, or the pension is less lucrative than promised or even the fund goes bankrupt. Heeding this is summed up in the slogan *carpe diem!* (Seize the day!) Except that doing so is selfish and thus allegedly immoral. The view of selfishness as immoral was questioned already in antiquity when the recommendation of enlightened self-interest (*eudaimonia*) became popular. The enlightenment that this slogan refers to, however, is problematic. The reference can be to common sense and it can be to science. In any case, when its application fails one can easily blame the failure on the insufficiency of the enlightenment involved rather than on the proposal itself. Spinoza suggested bravely that this is the right attitude regardless of its outcome, as it leads to the best possible outcome even when the best is not good enough and the only one that guarantees peace of mind regardless of the outcome. As things stand, his philosophy is the best available. This is not to say that its application is unproblematic. Spinoza himself applied it to politics in his *Tractatus Theologico-Politicus* that he probably gave up after the failure of his political involvement, as he wrote in his posthumous *Tractatus Politicus* that is but an exciting outline of a projected book.[3] Russell, the doyen of twentieth-century rationalist philosophy, was a Spinozist, although in a modification under the influence of Nietzsche who took the will to power as an irrational force, found in all individuals, that dominated all activities, regardless of their ends. Unlike Nietzsche, however, and in line with the philosophy of Spinoza, Russell preached the taming of our instincts rather than the yielding to them that Heidegger has preached. One way or another, here we see the failure of eudaimonism: it was the idea that, as Hume has put it, one can be trusted to take care of one's own interest better than anyone else. Notice that this wording does not refer to self-knowledge but to trust or rather to its failure: self-trust fails less than trust of others. Nor does it consider morality one way or another. This version of utilitarianism is minimal and so it is best. In particular, the initial preference of utilitarianism was the wish to eliminate morality altogether, which is a mistake. Rather, its force is the motive behind it, namely, the wish to eliminate preaching (mainly but not only due to hypocrisy as its frequent motive). And so, the best version of utilitarianism is the idea that we should act in our enlightened self-interest as best we can while being self-critical so as to improve.

The hope is to behave properly. We do not always know what our enlightened self-interest is: how does this translate into conduct? That depends on current circumstances, and we do not always know what the current circumstances are. Even when we think that we do, some unknown facts may be relevant to our decision problem, so that we may want to familiarize ourselves with the situation better before we act. Now the decision to study the circumstances in which we find ourselves may be just an excuse for procrastination that stems from fear of taking responsibility. How much study is it reasonable to undertake before action? Some books of rules, some traditional texts, come to answer this question. Ideally, they list all possible sets of circumstances and offer prescriptions for each of them. Yet most people are not familiar with the books of rules. Moreover, new circumstances turn up regularly. This is precisely why there are studies of ethics, of the rules by which to prescribe rules of conduct that prescribe the proper conduct under given circumstances. Yet the rules of ethics may ignore some circumstances since these are impossible, yet in time they may turn up as possible or become possible due to the invention of new techniques. Indeed, some patterns of behavior are obviously proper and some obviously improper; but there is a broad gray area where the rules, whatever they are, are not obvious. When devising rules for these we need criteria of what rule is proper what not. Such criteria are hard to come by. To that end we try to choose only the most significant and as obvious as possible.

As it happens, there is one simple rule that most writers in the modern civilized world take for granted. It is the rule that we must respect human lives as such. As Kant has put it, we should never consider a human being a mere means since every human being is a value; it is an end in itself. Shaw objected. He wrote in his preface to his 1904 play, *Doctor's Dilemma,* that this is practically impossible.[4] People may invest all their resources, he observed, in efforts to prolong their lives, and unconditionally so. Society, however, cannot behave this way: it cannot afford this expenditure. Before the collapse of the Soviet Union and its satellites, their capital cities were known for their drab appearances. Their apologists offered a general explanation for this: the cost of cosmetics is better spent on health. This turned out to be a lie: the communist leaderships did not care for the lives of their citizens. Thus, the Soviet Union refused to pay for medication for its average citizens after they had reached the age of 60 years. Yet the question remains: is it wise to spend on cosmetics money that could be used to save lives? Surely individuals facing this question will give up the use of cosmetics to prolong their own lives, but should a society, or rather a state, not do this? As long as we have no answer to this question, our views on public expenditure are defective.

Now no doubt, our views on public expenditure are defective. There is no investment more profitable than in education, yet budget cuts usually apply to the expenditure on education first! Still, in order to remedy this outrageous situation, we need some rules for allocation of public monies, including rules for expenditure on cosmetics. And if lives are really sacred as we say they are, how dare we spend on cosmetics when we are short of money for, say, cancer research? What are the proper rules for the allocation of public moneys?

We can learn something from the sad experience of socialist states: saving on cosmetics does not allow for the raising of the budget for research. Indeed, this is known as Keynesian economics: under some conditions spending on luxuries is just the right way to prevent economic crises and the agonies that they incur. Now possibly we live under conditions different from those that Keynes was referring to. Or he may have been in error. Yet this is obvious: we do not know how markets behave, and there are some conditions under which saving on cosmetics is bad for the economy. Not only that. When Britain introduced laws that facilitate purchase by instalments there was some public protest. By puritanic ethics, living beyond one's means is immoral, and deferred payments are obviously that. It took some effort to convince the British public that allowing for an economic crisis is morally much worse. The critics of Keynesian economics, the advocates of free market, objected to his proposal for governments to intervene in the market fiscally (by means of deficit budgets). Their theory led to advocating no intervention in the market, yet they advocated monetary intervention (roughly government trading in bonds or in foreign currency). This enables them to use their personal judgment without the use of a theory for recommending government activities. And then they cannot learn from their mistakes. To me it seems obvious that such conduct is irresponsible and thus immoral. I could not find any text on morality that says this. Come to think of it, most texts on morality discusses private affairs, since morality concerns individual conduct; the first to discuss public morality—the rule of individual conduct in public affairs—were Russell and Popper. This is odd, since the violation of these rules are well known as corruption, chiefly the bribery that already Holy Scriptures have proscribed.[5] Plato's *Politeia* too encourages this discussion, as it presented politics as ethics writ large[6] yet tradition has managed to ignore these cases. To be precise, Plato discusses justice, but with no distinction between the moral and the legal sense of the word, between individual and legal avoidance of discrimination. Now clearly the discrimination that violates the law is different from the one that the law sanctifies. The latter invites legal reform; what the former requires is not that simple. Moreover, there are conditions for legal reform. Legislators in the United States recorded a few times legal discrimination that they felt they should render illegal without doing so in consideration for the hostility that public opinion showed toward such reform.

Hence, legislation requires education and vice versa. This is a vicious circle that legislators hope to overcome stepwise.

It is hard to say more about this in general, but one can observe that legal reform is in strong demand when upright people do not allow themselves to behave properly since the law makes them pay dearly for it while reducing the benefit that it may cause. Yet even in such situations individual conduct in a supererogatory manner may have a great educational effect that may help introduce the needed legal reform. This is particularly the case when conduct recognized as admissible does not live up to the level of minimal legal demands.

NOTES

1. Karl Popper, *The Open Society and Its Enemies*, 1945, vol. I.
2. *Genesis* 3:19.
3. Benedictus de Spinoza, *The Collected Works of Spinoza*, vol. 2, 2016.
4. Bernard Shaw, "The Doctor's Dilemma," in *The Doctor's Dilemma, Getting Married, and The Shewing-up of Blanco Posnet*, 1911 (1904).
5. *Deuteronomy* 16:19.
6. *Politeia*, 369a–b.

Chapter 11

Minimal Standards in Politics

Some people find it shocking that minimal moral standards differ from political ones. They are in error. At the very least minimal standards should guarantee some reasonable quality of life at least as long as there is no emergency. The guarantee cannot be absolute, but at least it should hold under normal circumstances, whatever these are.

The paradigm case of the difference between moral and legal standards is the case of lying. It is decidedly immoral and as uncontroversial—except for the extreme cases generally recognized as possibly mitigating circumstances. What these are we need not discuss here. Suffice it to notice that (as the default option or as the first approximation) lying is proscribed when it is done with the intent to deceive (to exclude the hypocrisies commonly taken for granted). The liberal legal attitude to this is different. In all liberal countries the law proscribes lying only in two specific cases. The first is lying under oath since otherwise promises to speak the truth would be possibly as empty as any other promises. Also, lying is illegal when done with the intent to defraud, in which case the evil that the law combats is the swindling not whatever oath that goes with it. Many people suggest that lying should be illegal. This is highly problematic. On the whole, the wish to have the law perfect is practically impossible, so that the aim of the legal system is just to eliminate the worst cases, and even those only as best as possible, of course.

This idea, common sense as it is, conflicts with utopianism, which is a very popular idea. Utopians begin their inquiry with the question, what legal system do we want? Or, what is the best legal system? Once we know this, we can discuss the question, how, if at all, can we implement it? Or, alternatively, what is the implementable system nearest to it? It is far from obvious what is the better policy, to seek the weakest points in the current system or to seek the best system that we may try to implement. Perhaps the best is to try both. Perhaps, because this proposal may be dangerous. The philosopher best known for this warning is Popper. One thesis of his *The Open Society and Its Enemies* is, attempts to create heaven on earth turns earth into hell.

This proposal may lead to the proposal to seek a minimal political system. Such a proposal stands on its own, of courses: Friedrich von Hayek proposed it as the initial step. He supported both democracy and the idea of an economy with an utterly free market. He faced the hypothetical question: which system will become faster into a democracy with a free market. A society with a democracy or a system with a free market? He said, a free market will lead to democracy faster than a democracy will lead to a free market. This raises a simple question: what legal system is best for a market that is absolutely free? For, obviously, in order to have a free market a society must respect private property, and this respect is expressible in diverse legal system. An example is the question, does such a market need truth in advertising? As it happens, when the legislation that requires truth in advertising was contemplated, free market advocates opposed it on the ground that the free market imposes it better than the government. On the whole, what the free market can do the government should not do, as the government needs bureaucrats and these may undergo corruptions of all sorts. Now the free-market advocates tend to oppose all legislation that is intended to protect the interest of the individual if this protection can be left in the hands of the individual. The assertion that too many individuals are not capable to protect themselves so that there is a need to protect them is known as paternalism. The strongest argument against paternalism is not that individuals know what is good for them but that they can be better trusted to take care of their interests than the state (Hume). This is very nice, except that it is not clear how we should implement it. For example, should health insurance be obligatory? Should motor-car insurance be obligatory? In many modern countries, health insurance is not obligatory and motor-car insurance is. Why? There are many questions of this sort. The most important one concerns compulsory education. Even extreme Hayek-style liberals demand compulsory education. The writers who oppose it usually oppose all schooling. The proposal to have compulsory schooling with no compulsory education is the most unpopular even though it is the most reasonable under current situations.

Thus, the idea that we want minimal government is far from clear, yet the most popular version of liberalism is the proposal that the government should act like a night-watchman. We have seen that in fact this is not a sufficiently detailed proposal, but this fact is hardly a criticism: it leaves the details to those who wish to implement it. For, whatever minimal government is, it is not paternalist. This suffices to show that it is very hard to implement the idea of minimal government, since some paternalist laws are hard to eliminate—in democracies, not to mention anti-democracies. Compulsory education is one. So-called victimless crimes are others. These are self-harming activities, such as addictions, especially gambling and the use of mind-altering drugs. The claim against them that is not paternalist is the claim that they harm society excessively. This claim is testable. Those who participate in the debate should test it. They do not, since most of the advocates of decriminalization of drug use support it and those who oppose their decriminalization oppose drug use. The liberal idea that self-damaging actions should be legally permissible is unpopular, for the reason that most individuals prone to self-damage are infantile or that this action endangers society. In cases in which these valid arguments hold, governments should initiate some educational programs to alter the situation. They do not. This shows how important is the discussion of minimal standards of politics: to implement these society must achieve a minimal level of public education. This is not surprising: whatever democracy is, it is hardly possible to implement it in a society without a certain level of liberal education. To be liberal, a society must be tolerant, and toleration requires a level of education that is not sufficiently widespread. To date, over eighty percent of the world's adult population are illiterate! Now clearly illiteracy is the most significant contributor to poverty and the desire to avoid poverty is strong. Yet the obstacles to it are difficult to overcome. The two most significant ones are the scarcity of teachers and intolerance. Now the question what to do about this is too difficult to discuss here, and some international agencies are working on it. So, let us ignore it here and assume that we discuss only literate populations, and even democratic ones.

Hence, the discussion of the question, what are the minimal standards in politics, leads to the discussion of the question, what are the minimal standards for democracy, and the discussion of this question leads to the discussion of the question, what are the minimal standards for toleration.

The answer to this question is found in Popper's already mentioned *The Open Society and Its Enemies*: the toleration and even the encouragement of criticism. A society that tolerates criticism, then, is tolerant. This is not to say that a society that is tolerant encourages criticism. Popper has mentioned the fact that commercial societies are tolerant but seldom encourage criticism. All this is in very broad outline. We may go into some detail here.

Some societies tolerate criticism with some kind of limitation. The most common limitation on criticism is the exclusion of principles: they are exempt from criticism or above criticism, says a popular slogan attributed to Aristotle. This of course makes criticism hardly of any value. First, because every criticism of any item may turn out to endanger the principles. Indeed, this was the idea behind Plato's proscription of all criticism. Even he was taken aback by this sweeping proscription; for, he allowed some marginal exceptions.[1]

The distinction here is less sharp than it looks. The grey area between being open to criticism and being immune to it is that of hypocrisy: it is never clear how much what we say we mean and how much we say is mere lip-service. Indeed, much that we intend to say as mere lip-service turns out to be meant literally, whether as what we think or as lies. At times it is unclear as it is a kind of self-deception, as Nietzsche has noted. He wisely said, lying is unwise as it leads to self-deception. A particularly vicious and common kind of self-deception is what begins as permission to politicians and to generals to lie. This permission is most damaging, yet it is hard to do anything against it. Great efforts to avoid it are dangerous in a different way. This is what allowed an inveterate liar to accede to the presidency of the United States of America. It is by a dint of good luck that the damage he caused did not amount to a catastrophe. Perhaps this will teach Americans to appreciate truthfulness—such as that of Abraham Lincoln—a bit more than they do now. But other than advocate the demand for the respect that politicians should show to the truth, there is little that one can do about it, especially in the United States of America given its populist tradition. This is particularly due to the difficulty to differentiate innocent error from downright lie. It was William Hazlitt who said, it does not matter whether the wrongdoer is a knave or a fool, since behind a fool there is always a knave. This leaves honest error in the grey area, since its source may be negligence, and negligence may be admissible and alternatively it may be culpable—especially the error of electing irresponsible leaders. Admittedly, as Popper has suggested, the aim of democratic elections is to undo past errors; yet for this idea to be operative, sufficiently many voters should care for the future of their country in sufficient critical mindedness. This is far from easy even in the best of situations. My own compatriots are captives of a myth that prevents them from preferring civil society over community, thus preventing my country from entering the modern world, even though other than politically Israel is quite modern. This creates tension that is felt in everyday situations. It is thus very hard to know whether Israeli politicians are knaves or fools, although it seems clear that sufficiently many Israeli citizens foolishly refuse to have their politics normalized.

It is thus of a very personal interest to me to find out how to help effect the transition from communal life to civil life. The conspicuous difference between the two concerns the minority population. In communal societies, minorities suffer political discrimination; a civil society is supposed to rule this out. Even political scientists, who are supposed to display disinterest toward this, even they often fail to see that a normal, healthy society is both communal and civil, where cultural communities discriminate between members and non-members but without allowing any civil or political discrimination. What happens in most countries is that they deem themselves either communal or civil. The wonderful exception is the United States of America where both civil society and community are greatly appreciated and where therefore the separation between state and church is so significant and operative in daily life. Many thinkers consider this the separation between politics and religion. This is absurd: religion plays a significant role in politics and so religious thought cannot ignore politics. The same holds for politics: politicians cannot ignore religion. The separation is merely between the leaderships of the state and the churches. The trouble stems from the confusion of the distinction between politics and religion with the distinction between political and religious leaderships. The difference between them is conspicuous since politics and religion cannot be separated but their leaderships can be. In civil societies they are.

To be precise, the separation between church and state is not universally practiced in the modern liberal world. The alternative to it is what Spinoza has proposed: the church can be a part of the state. In Scandinavia this is the practice, and it works: political discrimination by religious affiliation is minimal there. Clearly, it matters little how discrimination by religious affiliation is prevented; it is its prevention that toleration requires, and it is toleration that belongs to the minimal requirement from civilized politics.

Yet, clearly, toleration is not enough. The idea that it is belongs to classical social philosophy that presented people as inherently good. Perhaps this was a vital stage in the development of Western social philosophy, since until then it was permeated with the opposite view, the theory of humans as evil, the doctrine of the Beast in Man, of the Original Sin. Indeed, the doctrine of humans as good is very uncommon. The default option is that of Original Sin. Now this idea is often linked with sex. The *Book of Genesis* report that Adam and Eve become aware of their nakedness only after they ate from the Tree of Knowledge. (They then used fig-leaves to cover their genitals.) This doctrine is additional: we are evil because we are captives of our (sexual) appetite: we are evil because we are captives of our greed. This idea is popular for pragmatic reasons. For, a basic item in all known cultures is the view that individual and society are in an irresolvable conflict: society wants the individual to overcome greed and laziness so as to be frugal and industrious. This idea translates into the work ethic that is very popular although it is pernicious and should be eradicated.

This proposal—to eliminate the work ethic—evokes a strong negative response. This is so due to the assumption that people hate work and would rather do nothing than work. This idea is one empirically refuted in a very clear manner, yet it still is very popular. It rests on a more general folly: without fear of hell people will not be law-abiding. When religious people learn that their agnostic neighbors are better citizens than they are, they express incredulity or incomprehension. This seems to me very significant. Opposing it without opposing the work ethic is futile and even inconsistent. The situation here is confused because we have to make it clear that the hatred of work is right and yet the work ethic is wrong, because we can create jobs that are coveted. No doubt about the following thesis, and no disagreement with it.

Many people who hate their jobs would love to work hard as physicians were it at all possible. Hence it is not work that they hate but their jobs.

The recognition of this obvious yet unpopular idea that some jobs are desirable leads to the question, what job is desirable, and can we eliminate all jobs that are not? Consider the fact that we fear the introduction of robots as the creation of unemployment whereas we should rejoice at the elimination of undesirable jobs from the job market. The reason we do not is obvious: the profit from the introduction of robots is not distributed rationally. True, the inventors of the robots and the entrepreneurs who introduce it must be compensated, and even generously so. Yet not at the expense of others. To do so some apt legislation is required. The trouble is in the dominant ideas about legislation, as well as the current practice of legislation. This is patently at fault. Hence, those who care about public affairs and who are willing to do something about it, should study the theories of legislation, criticize them, and then try to improve them. This requires improvement of public discussions in general. That is to say, procedures for public discussions should be discussed publicly. This is easier said than done.

Those who are inactive because they find the task impossible should note that much progress took place in recent years. We need not go far to notice that this is so, since dictators invest much effort in propaganda, namely in getting their message to you. And so, you may notice with ease that in any country, regardless of its regime, the more educated its people, the less prone its politicians to lie. For, education is the best tool against credulity—which means proper education, which means education that includes training in criticism. What democracy needs, then, is a critically competent people. The more critical competence, the more democracy and truthfulness. Hence, there is no need to legislate for truthfulness: it is much better to legislate for the availability of good education: citizens should not be deprived of the education they want for lack of money. The simplest way to reach this is for all education to be free. People repeatedly object that there is not enough money for that. This argument is always valid and in principle so: no matter how much education is available, it is insufficient. Yet there is some specific information behind this argument, and it should be made explicit, since it is common sense but erroneous nevertheless. It is that the normal average modern state cannot afford free university education. In truth the improved university will cost much less than the current one. That this is easy to do is clear: not so long ago—since the end of World War II—the university was much more efficient than today.

What made the university so very inefficient was the ability of some professors, the physicists, to be precise, to be able to contribute to armaments new weapons of mega-kill: nuclear weapons. When the Pentagon was unable to mobilize the arch-physicist J. Robert Oppenheimer, they put him on trial, presumably in order to shame him. In this they failed: his social position only improved as a result. (He became head of the prestigious Princeton Institute for Advanced Study.) The Pentagon decided then to mobilize the academic administration instead: they made a rule that all grants that academics received should be supplemented by equal grants to the administrations of their institutions. Grants to unattached individuals were stopped; even artists had to become academics and even art institutes had to become parts of the academic system in order to receive the coveted monies. The university administrative staff that was initially minute became overnight at least as large as the academic staff. This we can easily reverse by returning to old practices. For example, the administration sees to it that a student will not register to a lecture course in Algebra II before passing proper exams in Algebra I. Moreover, students who took exams in Algebra I will not pass it, no matter how knowledgeable they proved to be, unless they were registered in an Algebra I course. Very important. More important, students used to carry with them their records: what courses they took and if the courses were accompanied with exams, how well they did in those exams. For a four-year course that led to a bachelor's degree they had to take three or four exams. Today in each semester a student has to take a few exams. This does not mean that academic education is better now than it was traditionally.

The great difficulty in deciding what is the minimal standard for efficient academic education and how to implement them should be indicative as to the difficulty in deciding what is the minimal standard for efficient political systems. Where then do we go from here? We may despair of reforming academe or politics or anything else. Despair is always the wise default option. Indeed, the idea that we should behave this way is conservative politics and it is indeed the default option for almost all politics—for all political systems except those that advocated the idea that science should help improve politics. The exception was the system known as radicalism. The radicals wanted to destroy the traditional system. (The word derives from the word "radix" that means root, to designate the recommendation to uproot the old before implementing the new.) This is how the introduction of the idea of science to politics led to the French Revolution. Although that revolution was obviously a failure—it led to terror and wars—it still is a symbol for enlightenment! Parliaments practice piecemeal reforms and most modern countries are parliamentary democracies. Nevertheless, when in 1945 Popper recommended this in *The Open Society and Its Enemies* as a model for rational politics, he was derided and called a cold warrior.

To backtrack, in search for a minimal standard for politics, viewing despair as not an option and rejecting radicalism as impossible, we may still opt for near-radicalism, except that Popper's demand that we try to avoid errors that cause irreversible damage opposes this too, except for desperate situations.

It seems that the answer is to be found in Western-style liberal parliamentary democracy as practiced in Western-style liberal nation states. The question remains, what is so special about it? The best answer to this question is Popper's: the encouragement of criticism leads to learning from past mistakes and hopefully also to avoiding some possible mistakes that look reasonable enough yet fail to pass critical examination. To insure this to some extent (total insurance is impossible) we may develop traditions of concern for one's society that encourages some political involvement of some ordinary people in politics by legislating some simple incentives. These should lead to interest in public affairs; this interest plus the critical attitude should suffice to ensure a reasonable quality of life at least under normal circumstances.

It is generally agreed that a reasonable political regime should ensure reasonable standard of living. This standard varies, of course, so that with technological progress the standard of living rises, to reduce the average work load of ordinary citizens and to raise their standard of living, as well as to ensure health care, pension, and whatever else needs attention for securing the customary normal reasonable way of life. This looks trivial, but it is not. Some traditions enable citizens to take responsibilities and perform their life choices reasonably well. Other traditions make responsibility cause personal stress. The ideal of the French Revolutions was the achievement for the average citizen of a life of autonomy and enlightenment. The failure of the Revolution led the French parliament to spend much time in deliberations on education—in efforts to make the average citizen morally autonomous. The Reaction to the Revolution declared average citizens unable to be autonomous and take responsibility for their life decisions. It called for the abolition of civil society and the return to community life. The choice between civil society and traditional community filled the socio-political literature up to World War II. The advocates of community declared society a collection of alienated individuals, unable to interact as they do in an integrated community. It seems that by now consensus is being reached on the matter: modern life requires both civil society to which all citizens belong and diverse communities to which they may choose to belong. This duality does not ensure satisfactory life for all, as this requires support for both autonomy and dependence on tradition, as well as the ability to decide whether to integrate into society or into community or into both.

NOTE

1. Karl Popper, *The Open Society and Its Enemies*, 1945, Ch. 7, n. 20.

Chapter 12

Minimal Standards in Science

Science is unique: wherever you find science, the search for its source leads to Greece—to the presocratic thinkers, to be precise, or even to the first of them, to Thales of Miletus. What was specific to him? He travelled, he was learned, he had some achievements, displayed in the learned or empirical ideas that tradition ascribes to him. (His writings are lost.) These were not unique. Indeed, both Egypt and Mesopotamia were blessed with knowledge way superior to that of other cultures among their contemporaries. This made Derek J. de Sola Price declare Mesopotamia the cradle of civilization.

Few historians of science or of culture share his view on this matter.

Consider Joseph Needham's monumental *Science and Civilization in China*.[1] Needham struggled with the absence of a scientific tradition prior to the arrival of Father Matteo Ricci in China in the early seventeenth century. Before that China boasted quite a number of scientific achievements, say in physics as displayed in bridge-building and in astronomy as displayed in still extant records, yet science it did not possess. What was missing? Price could not say, since he identified science with scientific achievements. This is *a priori* impossible: all known human settlements, throughout history, had significant scientific discoveries, beginning with the discovery of the control and use of fire as well as of the weapons and of the wheel, not to mention the building of huge houses and long bridges and even colossal dams. All these monuments display the possession of great funds of knowledge. Ones that can come only with traditions of accumulation of knowledge.

Historians of science do not discuss institutions. Attention to these came with the writings of Harry Austryn Wolfson (1887–1974). He viewed the three Western Abrahamic religions as variants of one, yet he divided this single tradition both philosophically—methodologically, to be precise—and institutionally to two. He said that both traditions were created by Philo Judaeus or Philo of Alexandria and terminated by Spinoza.[2] Now methodologically the division was between rationalists and mystics; institutionally, rationalists were housed in institutions of learning like monasteries and universities.

(Both types were church institutions initially and until their secularization that the French Revolution imposed.) Mystics were loners, often hermits living in the wilderness, and entering society at large, if at all, when they join society intermittently, usually in courts, so that they were identified as courtiers. Thus, Tycho Brahe, William Harvey, Johannes Kepler and Galileo Galilei were all courtiers. The Copernican Revolution was both astronomical and methodological. In the preface to his book (in the letter dedicatory to Pope Paul III), he said, since the Ancients disagreed among themselves, they cannot serve as authorities. Medieval thinking was conciliatory: it included methods of removing disagreements. Medieval thinking is best characterized by the title of the major and most influential book of Al-Farabi, titled *The Philosophy of Plato and Aristotle* that was often labelled as *Plato and Aristotle Never Disagreed.*[3]

All this is beyond my comprehension. All students of skepticism, from Plato and Pyrrho to Richard Popkin, they all agree explicitly that doubt paralyses. Now imagine a platoon of soldiers after a battle finding themselves trapped behind enemy lines. They must act and they have no idea how to. They may be paralyzed, but they need not be; they may deliberate, decide intelligent or unintelligently, and act. This refutes the thesis that doubt paralyzes. Nor is this the only refutation. Notoriously, a change of leadership of an organization for the better often means that the new leadership is more decisive. Its conduct is another refutation of the thesis in question.

More generally—much more generally—we wish life to include learning of all sorts; some of these we lump under the title of learning by trial and error. How is it that both the slogan that doubt paralyzes, and that learning is by trial and error are both popular is a miracle. Perhaps because the one slogan fits practitioner and the other fits researchers. This brings us back to Wolfson's division of all intellectuals to two: academics-rationalists and loners-mystics. And then Boyle broke this dichotomy when he created the Royal Society of London as a collection of amateurs who are economically independent and so belong to no economic supporting institution, and in principle so.

The idea behind modern science, and it is not clear how much this is thanks to Boyle's idea of amateur science, is that science needs a tradition, but one that is not intellectually binding. How powerful traditions are can be seen from the fact that traditional European folk medicine travelled to the New World together with the early European migrants, seemingly with no written text yet with so much information left intact.

The same we can say of Thales of Miletus. What then was specific to him if not the specific scientific information and techniques that tradition ascribes to him, then it is his starting a new kind of tradition, the one that led to the rise of (Greek) science. What was that kind? What is (Greek) science? Discussions of this question take their cue from the obviously specific to Greek science: Euclidean geometry. What is peculiar to it is clear: it uses the axiomatic method, that, all results are deduced from simple set of axioms using intuitive logic. Less specifically, it is its deductive method. Logic is present in all cultures. It is hard to find any long text with no deduction. Yet systematic deduction like the ones used in Euclid is rare indeed. (The texts of Archimedes were lost, to be rediscovered on palimpsests around 1900.) Not in vain was Euclid a best seller for centuries second only to the Holy Bible. Our question has thus become more specific: did Thales contribute to Euclid in any way? If so, how? The received answer is interesting. It is the idea that all scientific explanations rest on the assumption of naturalism, the replacement of mythical explanations by origin with natural explanations by assumptions of natural laws. The systematic effort at explanation of laws by higher-level laws required the systematic effort to use deductions. Was Thales responsible for this? Or at least for a part of this? As tradition ascribes to him naturalism, perhaps the answer is in the affirmative. We do not know. Yet the question is, what is the advantage of naturalism over its predecessor? The answer sought for is, logic: naturalism has the advantage of the systematic use of logic. How? Take a mythical explanation: something is the case because something else willed it. The minimal change required in order to render it scientific is expanding it: something is the case because something willed it and also that something else had the power to make it so. This then can be reduced: the will could be eliminated while the power remains. This is where science is more abstract than myth: the laws of nature want no force to enact them. (Kepler still postulated angels using their own power to rotate the heavenly wheels. Yet, in his theory the angels could be forgotten, as indeed they were.)

This, however, is insufficient. The myth world can survive the elimination of the subject willing to empower it, since it allows for wills hovering in space seeking a subject to hang on to. On this matter the myth world is more abstract than the world of science, not less. Naturalism eliminates this advantage too. Weber called this further elimination the disenchantment of the world; Erwin Schrödinger called it the Greek discovery of nature: the Greeks declared nature meaningless and this, according to Schrödinger, is naturalism, and the rise of naturalism was the rise of the natural sciences. This disenchantment of the world is what makes naturalism hard to swallow. A leading twentieth-century philosopher Jean-Paul Sartre still found the meaninglessness of nature hard to swallow and made this fact a cornerstone of his philosophy. (He had an unbelievably narrow conception of science.

His interest was naturally in psychology; he deemed behaviorism the only psychology that has the right to claim scientific status, yet he found Freudian psychology so much more appealing, and largely because of his personal experience: he was raised by his grandfather together with his mother, he reports, and he found her attractive.[4] Possibly comprehending this cost him great effort. For, he came to the amazingly silly conclusion that self-awareness and honesty to oneself is the most important requirement from any individual. All this deliberation goes with forceful emotional overtones. It was for him then understandably difficult to take seriously the basic idea of modern science that the universe is meaningless. He did not see that for researchers the truth and commitment to it is anything but.

To return to Thales of Miletus, science deems nature meaningless and this is an aspect of science that he has introduced to Western culture and civilization. How if at all is that linked to the deductive character of Euclid? This question is best attacked by discussing our initial question: what is the novelty of Thales's contribution? In response to this question Popper has offered a revolutionary new idea. Schools of thought, he observed, come to preserve the purity of the school's doctrine. When this doctrine alters, he added, the school's leadership does it surreptitiously. For, change is a part of a response to criticism, a response that begins with the admission of the validity of the criticism. As Bacon has noted (around the year 1600), the leadership of a school is reluctant to admit criticism for fear of damage to its reputation. The first exception to this rule was the school that Thales founded, said Popper, as it traditionally and openly ascribes to his leading successors ideas different from his. No surreptitious change and no break from the master to create a new school! Social continuity with theoretical discontinuity appeared on the horizon.

Thales had the idea that water is the one substance which all matter comprises of. This is a complex of ideas. The first is the idea that there is appearance and reality. Now the fact that appearances may mislead is common knowledge: it is hard to be unaware of it. Yet the idea that all appearances, even the best tested ones, are mere surface, that boring deeply into the phenomena delivers different information, this is less obvious. Greek science is unique, but the idea of appearance and reality is not: there is one idea that is present in all cultures: mysticism. It is the dual idea that in the last resort all is one, and that the experience of mystical union and only it reveals this truth. It deserves notice that nowhere on earth was there any idea about the acquisition of true knowledge other than the idea of the mystic union. Its outcome has two kinds of versions, universal and particular. The universal idea is that all is one; the particular ideas are found in the diverse religions that have organized churches that function also as schools of thought, as institutions that aim at keeping the initial doctrine pure. The view that there are (demonstrable) truths by nature and (arbitrary) truths by convention—universal and conventional truths—and that only truths by nature are compulsory or obligatory to admit, this idea is particularly Greek and it rests in the foundation of classical (Western) science that we gave up already. Today we view science as the best putative truth, not necessarily troth by nature and not arbitrary truth by (mere) convention but something in the middle: most scientific researchers accept Einstein's verdict: (current) scientific truth is not the absolute truth: if we are lucky, we may expect a scientific revolution that would drive science forward.

Sadly, most philosophers of science are very backward. Despite avowed admiration for Einstein, rather than admit his view of scientific truths or at least put it on the agenda for public discussion, their public agenda concerns the credibility of science: why should I believe science? This question rests on the opinion that it is obligatory to believe science, an opinion that is clearly anti-scientific, since science opposes all intellectual authority including itself. Moreover, as Larry Laudon has observed, most philosophers of science can hardly believe science since most of them are scientifically ignorant. Indeed, it seems there is no individual anywhere in the world familiar with all up-to-date scientific theories so as to be able advocate credence toward them all, much less to deliberate on their quality.

Now, no matter how little science philosophers of science are exposed to, even elementary education suffices to give them a taste of scientific ideas and even of scientific thinking. Yet the deep-seated idea that science is certitude spreads confusion. The reason is that the inability to achieve certitude misleads many philosophers to search for certitude-surrogates, chiefly near-certitude. For, whereas certitude comes to guarantee the universality or objectivity that characterizes science best, near-certitude does not do that, since it is a quality of conviction, and conviction is eminently subjective. Efforts to overcome this unpleasant corollary led to the view that the justified conviction is objective, since its justification is. Efforts to spell out the idea led to the conjecture that justified conviction is probability, since the probability of a hypothesis may rise with the increase of the body of evidence that supports it (at least in the absence of evidence to the contrary). What this probability is philosophers were not clear about, but they assumed that it is a function (with theory and evidence as variables) that follows the rules of the calculus of probability.

This Popper has refuted. Suppose there are two theories and one observation report, such that the evidence raises the probability of one and lowers the probability of the other, yet the one with lowered probability is more probable than the other (due to its high initial probability). Hence, not initial probability but probability increase may be a measure of credibility of a hypothesis. Unfortunately, however, whenever infinitely many options are available, the probability of each option is zero, and then, no matter how many observations tally with the hypothesis in question, the probability that the next observation will tally with it is still naught. In brief, the idea that the calculus of probability will help us find a certitude-substitute is defunct. Sadly, too many philosophers of science find it too hard to let it go. Our question is, what is the minimal standard of science? To answer this, we wanted to characterize science. And we wanted to do so with no reference to what any individual feels about science, even if that individual is a *bona fide* scientific researcher. We tried out the idea that science is the body of theories that experience supports. This has raised the question, what is the empirical support for a scientific theory? We tried out the idea that it is probability. This does not work. A better idea is that empirical support for a theory makes its probability increase. This too does not work, since scientific thinking, Popper has observed, is bold, and the bold ideas are improbable. What then is the empirical support that renders an idea scientific?

Whewell who wrote in the middle of the nineteenth century answered this question. He rejected toe popular idea that science begins with indiscriminate gathering of empirical evidence. Rather, he said, it is the search for explanation of known empirical information and the test of the hypotheses invented to that end. Empirical tests of hypotheses are likely to show that we have failed to imagine the truth. And so, when a hypothesis is refuted, attempts to replace it are called for. This process is in principle endless. In fact, however, with luck and ingenuity a researcher hits upon a theory that experience agrees with. For example, Kepler imagined one obit for Mars after another and refuted them one after another until he hit on the right ellipse.

Whewell's theory was declared unsatisfactory and it was forgotten for quite a few decades. The reason it was unsatisfactory is that according to it scientific progress depends on luck and so it is not assured. The idea of scientific method that was generally received in his time was that of an assured procedure, an algorithm for success: here are the rules for proper conduct; if follow them, if you behave properly, then you are assured of success. Hence, if you have failed, you have no one to blame but yourself. This is the idea of full justice, of divine justice on earth! How come people fell for it? People are credulous, observed Samuel Butler. This is no explanation. Why do people believe that the future is rosy? Why not believe that the future is bleak? After all, there are so many more ways to suffering than to joy. Moreover, the view prevalent in almost all societies on this earth is that the individual—any individual—is prone to selfishness and laziness that only cause havoc. As recent liberal policies show, so much of what was required of the individual as necessary sacrifice has surprisingly turned out to be quite unnecessary. For the first time we are raising children with no threats of punishment constantly thrown in their faces, without subjecting them regularly to corporal punishment, often with no justice. Why then should we not hope to be able to create heaven on earth? Why not dream of peace on earth? The nuclear age, said Russell, opens to us the choice between hell on earth and utopia. For the first time, he observed, utopia became accessible. This was almost a century ago; was it refuted? Is his observation scientific? What renders an observation scientific? What are the minimal standards in science? We do not know. Discussions of this matter went on for centuries, and they seem far from reaching any conclusion. Science was defined as provability. And by this definition science does not exist and even cannot exist in principle. (The principle is that of fallibilism: nothing on this earth is perfect; nothing on this earth can be perfect.) Once it was claimed that Newtonian physics is perfect, that his theory of universal gravity is. This was not unreasonable, even though it was a mere hope.

Every attempt to refute it failed. Even skeptical Hume was impressed: The severest scrutiny, which Newton's theory has undergone, proceeded not from his own countrymen, but from foreigners; and if it can overcome the obstacles, which it meets with at present in all parts of Europe, it will probably go down triumphant to the latest posterity.[5]

The reference here to foreigners, incidentally, is mean to exclude the suggestion that the scrutiny was not as severe as possible due to nationalist bias. Hume could not imagine that the bias could be not national but human. And his verdict was generally acknowledged. Kant, the most systematic and pedantic commentator on science ever, did not question the verdict; he did not ask, how could Hume, the greatest critic of science of all times, conclude that Newtonian mechanics is science proper? He did not even stop to argue that Hume was in error, that contrary to his assertion, even the theory that underwent the utmost scrutiny was short of proof. He did not discuss scientific method beyond observing that scientific observations too are theory laden. The discussion of scrutiny in his works is marginal and reveals that it troubled him, and rightly so, since it competes with his philosophy. Rather, he rightly found it troubling that confirming observations should mean more than that they only reveal the theoretical input that had gone into them.

Hume's critique of science, so famous and important, concerned only epistemology (the theory of knowledge), not of methodology (the theory of the way to do research). Moreover, his central critique comprises the claim that causality is not observable. For, it is essential for a clear concept of law of nature: without it all empirical confirmations are wanting as they may well confirm descriptions of merely incidental circumstances. Indeed, this was his theory of knowledge: science describes the facts about researchers: they describe mere habits, not causal laws. Habits may rest on laws, but more likely they do not. They do, said Kant, because our idea of causality is no mere habit: it is unavoidable. Maimon said, this claim for necessity is a mere hypothesis. Kant was lost for a response. He stooped so low as to answer Maimon *ad hominem*: he called Maimon a parasite, adding that this is due to his being a Jew. Parasite or not, was his criticism valid? The jury was out on this. The verdict, however, is incontestable: Maimon's critique is right.

Einstein has rejected all the philosophies of science that he had encountered, but only Kant's he declared dangerous: he found in it a threat of complacency. (But then he, too, was a Jew.)

We have inadvertently hit upon a very important idea that surely belongs to the minimal standards in science: no *ad hominem*: no matter who said it, the questions about it should be impersonal: are given assertions relevant? If so, are they true? And so on. Although scientific ideas emerge from individual minds, their scientific treatment is impersonal. This, although it belongs to the minimum standards of science, need not be true: science too is no utopia. The leading twentieth-century sociologist of science Robert K. Merton has discussed one of the most popular anti-scientific ideas, the idea that who says something is relevant to our view of it, or at least to our readiness to pay attention to it. He christened this the Matthew Effect: "that hath shall be given." Still, we may assume that favoritism happens in science less than elsewhere and that the damage it incurs when it is operative is short-lived. We may say this generally: whatever we view as an essential aspect of science and a valuable characteristic of it, is very likely to be violated here or there, but less frequently than elsewhere, and with little damage. This sounds suspicious: It sound like a Kantian proof: it is necessary for science; therefore, it exists. Not so. Rather, it is necessary for science; therefore, as long as science exist it exists. Hence, lovers of science should pay attention to it and warn the commonwealth of learning against serious threats to it. For, there is no guarantee that science will survive.

More than that. There are always claims for scientific character that accompany some items that are fake. Should we ignore these claims? Or should we ignore the items too? It is hard to say, since most of us ignore many items that are *bona fide* scientific, simply due to the inability of any individual to pay attention to everything scientific. (The last person who had a claim to anything like it was Whewell, who lived in the middle of the nineteenth century: it was said that omniscience was his forte and his foible.) Since people ignore so much that is *bona fide* scientific, *a fortiori* they ignore much that is possibly fake—except when they are researchers and suspect that the possibly fake is an answer to their prayer. After all, even acknowledged experts often dismiss significant contributions to science. This is scarcely condemnable given that most claims for novelty are duds. Here minimal scientific standards, the topic of this chapter, can be of tremendous use. Ambitious science schools aim at helping their students to be able to read scientific texts; they do not teach them how to discriminate, often because they do not know how. Yet the rule is simple and reading the following few words will make one expert in this matter. When glancing at a paper one should ask, what problem does it address and am I ready to address it? It is this that makes glancing before reading very advisable: a paper that addresses no problem in particular or a problem that is not one's agenda right now, is better left unread, possibly set aside for reading on a rainy day. It is not easy to find the problem that moves an author. Baconian tradition encourages researchers to report their observations with no background information and no problem. Experts are well trained in scanning research papers before study, since the study of a paper is time consuming and in these days of publication pressure much irrelevancy and repetition reaches the learned press regularly. Minimal requirements that readers may set are just what the doctor ordered.

NOTES

1. Joseph Needham, *Science and Civilization in China*, Cambridge: Cambridge University Press, 1954–present.
2. Harry A. Wolfson, *Philo: Foundations of Religious Philosophy in Judaism, Christianity, and Islam*, 1947; *The Philosophy of Spinoza: Unfolding the Latent Process of His Reasoning*, 1934.
3. Al-Farabi, *The Philosophy of Plato and Aristotle*, 1969.
4. Jean-Paul Sartre, *The Words*, 1963.
5. David Hume, "Of the Rise and Progress of the Arts and Sciences," *Essays Moral, Political and Literary*, 2015 (1776).

Chapter 13

Optimism as an Imperative

The words "optimism" and "pessimism" are often in use, usually in reference to specific cases, when assessing the likelihood of success or failure of a specific project. One refers then to the estimate of its success as optimist and to the estimate of failure as pessimist. Individuals whose default estimates of projects tend to be favorable are considered optimists. Cautions investors tend to seek the advice of pessimist experts on a project before committing themselves to invest in it. These are two different kinds of optimism and pessimism, of attitudes specific to projects or kinds of project—political, psychological, religious and metaphysical—and attitudes to life and to God's world in general. The general ones may but need not rest on some idea of rationality. One may perhaps view other cases as irrational.

Alternatively, one may view them as psychologically given, say, as matters of temperament, possibly inborn. When they pertain to matters political, they may influence the lives of whole populations regardless of the dispositions of their individual members toward optimism or pessimism.

All this rests on the presumption that some rational assessment plays some role in the formation of the disposition in question. The preferability of a pessimistic attitude over an optimistic attitude then comes naturally: the world offers small windows of opportunity, and we usually miss them. Rationality should then appear as helping to take advantage of them. And then, whether optimism depends on some rationality or not, by common sense and common experience, optimism seems inferior to pessimism: disappointment occurs more often than satisfaction by far.

To the extent that people try to assess situations rationally, things look to them quite differently: there is a limit to the possibility of an optimistic or pessimistic disposition to influence rational assessments of current affairs. Some situations invite a choice between optimistic and pessimistic rational assessments so clearly, that no individual disposition will block them, much less overturn them: no matter how optimistically or pessimistically disposed a person may be, no matter how often one has expressed general optimistic or pessimistic faith in the recent past, if that person lives in a besieged city on the point of collapse due to starvation, for a conspicuous example, then under the pressure of the situation we should expect that person to become most pessimistic. Nevertheless, one's optimistic disposition can make a difference: it may make one seek some way to overturn the situation and make it more hopeful. The new way of looking at the situation may lead one to try harder to break away from bad luck. A besieged unit may gather its resources and courage and try its best to break the siege and succeed or die in the effort.

Today, when we face serious threats of global destruction of the ecosystem and of the human habitat, there is a certain lackadaisical attitude toward global politics that is genuinely frightening to the point of leading to the assessment that all is already lost. Nevertheless, we should grant the Good Lord the benefit of the doubt and try our best to alter the dreadful verdict. Perhaps in such situations the rational thing to do is to overcome all pessimism in an effort to seek some hope, quite contrary to the attitude of abandoning all hope adopted by the Jewish leadership, in Europe and more so outside Europe. With hindsight we can easily spot reasons for optimism in bleak situations; with foresight we should look for them.

All this should make it clear that optimism and pessimism are not simple matters of disposition; similarly, they are not simple matters of assessment of the situation; they comprise complex rational assessments and careful, responsible deliberations. Nevertheless, one may examine the matter of given psychological dispositions, their roots, and their possible contribution to a more reasonable assessment and conduct. The matter of psychology, the *a priori* disposition to be optimistic or pessimistic, is usually presumed to be given. it is supposed to be an innate disposition; it is also supposed to depend on given, alterable circumstances.

All this is questionable: it is increasingly difficult to decide what disposition is innate and what not, and it seems increasingly clear that we do not quite know even what we mean by the concept of innate disposition, at least when we apply it to animals other than insects and crustaceans, particularly when we apply it to human beings. For, the complex situation is this. An innate disposition is a native disposition to react to the environment in a given way. The lowest animals have stable reaction-patterns fairly indifferent to a broad range of variations of the environment and no clear pattern outside that range. Higher animals exhibit a remarkable plasticity, which is expressed by the fact that their innate reaction-patterns are versatile and exhibit different regularities in different environments, whether we consider the environment physical, biochemical, or even social. In the case of human beings, we agree that our dispositions are to react in a great variety of ways to our educational environments: even if the disposition to learn languages is innate, as some people say, clearly it is such that we react differently to different linguistic stimuli, and the reaction is not simply that of aping others but something else, which is still not known in the least and is under intensive study.

The question of the possible innateness of the disposition to be optimistic or pessimistic is even more complex than other possibly innate dispositions; even on the supposition that we are born with the disposition to be optimists or pessimists, the question remains, under what conditions is this *a priori* disposition retained? We do not know. Yet the matter of personal disposition is hardly deemed problematic, since we assume that we are plastic enough to handle matters of practical significance as not mere matters of taste: we assume that we can reject the view that the matter is merely one of taste not open to rational debate and not given to resolution; we may assume, to the contrary, that debating concrete problems may be made relatively insensitive to the greatest variability of personal disposition, in the sense that we may try to open the pessimist and the optimist to exchanges of opinions and to debating them rationally.

All this is applicable quite generally. People often say that artistic tastes are given and not subject to debate. This assertion is very popular despite the ubiquity of its refutations: the taste for high art is acquired, so that most people simply lack it. Hence every case of developed artistic taste refutes the assertion in question. More than that: the evidence that seemingly supports it is that the diversity of tastes for art in general or for an art form or for a school or a tradition within it, yet this diversity is a refutation, not a corroboration of the assertion that artistic tastes are not given to rational debate. What veils this fact is that the obvious argument against it refers to long and arduous processes that are not always successful. (Moreover, art education is so poorly implemented that its failures often generate intense dislike for art.)

The suggestion here is this. In every situation, we may put to rational debate the difference between optimistic and pessimistic assessments. This renders secondary the difference between optimistic and pessimistic attitudes to life.

This very assumption, it turns out, is the central rational ground for all rational optimism. Throughout Western history, rationalism led to optimism and irrationalism to pessimism' and rightly so, since the greatest cause for optimism is the hope that we may use our rational capacity to ameliorate matters. What is this rational capacity? What is its practical expression? This is a big question: its investigation comprises a whole field of study: methodology (from the word *hodos*, meaning "way," like the Chinese word *tau* in "Taoism").

Methodology divides into three schools; the intellectualist, the empiricist and the dialectical. The intellectualist school has full control over logic and mathematics (that includes much of computer science and analytic physics).

The empiricist school includes vast practical areas, from mining to aviation. Indeed, whatever practice one discusses in detail, no matter how much theory-based, the final touches are matters more empirical than theoretical. Outside the fields that are obviously controlled by intellectual or empirical considerations, most of our studies are dialectical, often labelled traditional or trial and error. We forget that before Pasteur and Lister inaugurated modern medicine (under the influence of Semmelweis), medicine was almost entirely a matter of trial and error. Yet its biggest error, its use of bloodletting as a panacea, was stopped only as the result of the contribution of Adolphe Quetelet to the improvement of statistics to the point of its applicability to problems of discrimination between the relevant and the irrelevant (1830). This comes to show that the method of trial and error is no method at all: a method is a set of rules to apply whenever a problem of a given sort wants a solution. Trial and error demand resourcefulness that is not always there for the asking.

But is the proposal to put any proposal to rational debate, is this very assumption rational? This question reverberates through the philosophical literature and the negative answer to it was used repeatedly by irresponsible, irrationalist political leaders and ideologues to defend the vilest of political attitudes. It led Popper to defend, in his classic *The Open Society and Its Enemies* (1945), the view that the rational attitude is not itself rational but it is the least irrational. William Bartley III criticized this. He led the debate away from the political to abstract or to metaphysical directions. Worse, it was a retreat to subjectivism. Yet the question that Bartley was tackling is real: can we assume a modicum of rationality that will enable us to debate matters and come to assessments more rational than the merely arbitrary or the rooted in mere psychology? If we assume that even on pessimistic assessments it is most responsible to try to be rational, to hope despite all discouragements, then we may proceed, at least for a limited time span, and see where the discussion may lead us. Often escape from dreadful fate appears late, and from unexpected directions. Let us then attend in this spirit to the question of the appraisal of the prospects for survival of a venture, of a person, of a people, or even of the human race as a whole: let us ignore optimist and pessimist dispositions and ask responsibly, how good are the prospects of survival and what can we do to enhance them? That there is nothing more rational than the effort to appraise prospects is so regularly admitted that it has a parable to illustrate it which is taught to toddlers: King Canute foolishly tried to order back the tide; he did this by royal command. Now the tide stands here for what is known in the relevant literature as a (historical) trend. A trend can be (mostly) improvement of some quality or its deterioration; it can also be its mere stabilization. The standard example of a trend is the rise or the fall of the Roman Empire.

Generally speaking, in Western cultures optimism is usually preferred to pessimism. This optimism was neither psychological nor practical. We know that Roman optimism was not psychological since the information about their personal dispositions makes it clear that some of them happened to be profoundly pessimistically inclined. Nor was their optimism practical, as they differed on the practical affairs of state and fervently argued about prospects. Hence, their optimism was of a metaphysical type, one that is impervious to inclination, mood, or practical appraisal. There is no strong criticism of such an optimism, or to the alternative to it that is pessimism. Of course, at times the metaphysical attitude may translate into a practical one, and still without inviting serious criticism. The practical attitude that this metaphysics may suggest should be open to critical debate.

For example, consider risks to the future of humanity, the four P's, the Pollution of the environment, the Proliferation of nuclear weapons, the Population explosion and the Poverty of more than half of the population of the world. It is all too easy to defend the appraisal that humanity is consigned to expeditious extinction on account of the current situation, particularly the inability to face global problems and to control the four P's.

Nevertheless, this assertion is not scientific: it does not follow from any known scientific doctrine and is not scientific by itself (since science deals only with regularities). This is not to deny the possibility of a rational discussion of this matter and of some other matters raised by metaphysical doctrines, nor of the debate being profitable. It is merely to affirm people's right not to defend their views—this, obviously on the condition that they have no right to impose them either. But then, there is no right ever to impose any opinion on anyone. This is of course broadly recognized, even though it is contrary to the immensely popular view of Thomas S. Kuhn: he did declare obligatory faith in the current scientific paradigm, namely, in the metaphysical ideas that the scientific leadership currently endorses. But this he put as a condition for staying a professional scientist.[1] Moreover, since Einstein and Planck are obvious counterexamples to Kuhn's demand, we may ignore his constraint on the freedom of thought within science, his immense popularity notwithstanding. Rather, it would be wise to notice that what is *de rigueur* is only the admission of repeatedly repeated factual information that is declared repeatable. To this some commentators add, possibly rightly, the admission of the claims for high explanatory power of a theory, for the high degree of corroboration of a theory and for the refutation of a theory—always while admitting that the obligation to recognize factual information is limited to what is reported repeatedly and declared repeatable.

This is not all there is to the matter: what is important is not only what is *de rigueur* to admit but also what is agreed upon as the agenda for discussion: to the extent that science is a rational debate, it takes different people to agree on the agenda; to the extent that science is a public concern, it has a publicly set agenda. Who decides the agenda and how? These are questions that hardly anyone has debated, and for the following simple reason. The classical philosophers of science assumed that the agenda is the most rational one, so that there is no problem about it; contemporary sociologists of science are either in agreement with this or declare that the decision on the agenda is in the hands of the recognized scientific leadership. Who are they? Who are the recognized scientific leadership? The default answer is, the recognized scientific leadership are the researchers whose opinions are most popular in the commonwealth of learning. Who are these? We do not know, but we do know that astrologers are not and that recognized leading physicists are. Why? Astrologers do not count since their attitude to information is devious: it must be since too much easily available information defied astrology: the fates of individuals who share astral information vary.

We are back to the starting point of the present discussion. Different respectable traditional attitudes lead to different views on universal optimism; facts on the matter are too diverse to set to a plain picture. They concern two items. The first relates to future situations, the second to ways and means. Optimism and pessimism concern future situations on the tacit assumption that we are agreed about how to assess situations and about the assessment of current affairs. Naïve candidates for election are usually optimists: they say, we must win since truth and justice are with us: in the end truth and justice must win; you cannot fool all the people all the time. Moreover, we need not agree about ways and means: suffice that we agree that truth and justice will prevail.

The scholarly aspect of all this is easier to record and discuss on the supposition that we have a scholarly tradition that can be relied on when sifting the grain from the chaff. Now scholarly optimism easily spreads to become optimism in general. This merits stress: there is no clear dividing line between scholarly and popular literature since many able scholars sing to the gallery. Moreover, within philosophy the trend or tradition fosters and inculcates myths; this fact need not be taken too seriously, as one may spend too much time on it and forget the purpose at hand, which is to discuss the question, how serious are the theories behind pessimism and optimism in general.

The doctrine of historical trends comes from two interests, the wish to assess the current situation properly and the wish to develop a broader perspective. That is to say, there is a difference in trends, as there are local and global ones; the global ones are called historical. The example presented above, of the rise and fall of the Roman Empire, is from the writings of the German Romantic philosopher, Hegel. He assumed that there are global trends, and that the rise and fall of empires is a global matter, almost by definition: empires are global domains, not merely local ones. The discussion of the difference between the global and the local, to repeat, is a matter of a desire to have some perspective, and perspective imposes a new kind of discipline, very different from the scientific one. Scientific discipline is a matter of precision, and everyone who can be trained to prepare an experiment carefully and attend to detail is capable of scientific training. Not so the matter of a sense of proportion: it cannot be taught even if its growth may, perhaps, be encouraged. A sense of proportion even may conflict with the scientific attitude: a sense of proportion may require that we overlook an insignificant detail yet, as a well-known myth about science decrees, science rests on the demand that no detail be overlooked. All this is Hegelian lore carefully stated. The right attitude is very different: science comprises problems, conjectured solutions to them, and empirical tests of these solutions.

Since many Hegelians are not ready to stoop to careful wording, it may be advisable to be careful and say that there is no canonic wording in the Hegelian literature. I am following here the suggestion of my teacher, Popper, to try to present a doctrine before attacking it, and do so as well and as convincingly as possible. Let us agree: there is no reason why we cannot develop theories that overlook details. Even in mathematics this can be done: imagine what would have happened if Berkeley's valid criticism of the calculus were endorsed and the calculus would have waited till it was successfully taken care of. Moreover, there is even a mathematical quasi-justification of the oversight of details: the first great deviation from the traditional view of mathematics as algebra and geometry, the theory of groups, starts with the study of a structure that an operation on its members imposes on them in disregard for other properties of the group. This was an indication that numbers too are properties of sets regardless of its properties. Consider the formula "5 things plus 7 things equal 12 things" as long as the word "things" in it is fixed, it holds; not otherwise, as John Stuart Mill argued, noticing that a drop of water plus another drop of water may unite to one drop of water. Yet it is bigger than the initial two.

This leaves open the very interesting and practical question, Can people be taught to have a sense of proportion? If yes, can this be done systematically? This matter is under dispute. Unfortunately, the prejudice that determines the answerers to these questions concerns science itself: once we decide whether we advocate science or not, we are ready to say that science can or cannot help develop a sense of proportion, seldom the other way around.

There is no reason in the world why we cannot try to develop a theory of the sense of proportion and examine it empirically. The question is, will this attempt be successful enough to help us teach everyone a sense of proportion? Not only is the matter of a sense of proportion put in the wrong order, as instead of asking whether science can help the development of a sense of proportion and then decide to appreciate science or not, we express our appreciation or its absence by deciding a priori what science can do; we also decide about global trends and the decide whether recognizing them is or is not a matter of a sense of proportion.

The discussion of the question at hand has run wild: the question may be put in different ways. How much criticism may be overlooked in the name of a sense of proportion? How much should we ignore details in order not to lose sight of the general? The advocates of science and/or of criticism are reasonably motivated by the consideration that license to ignore details is license for dogmatism. The advocates of a sense of proportion, however, fear that the details, especially the criticism, will destroy all and leave us blinded and desperate in the shambles that it leaves behind. Clearly, both attitudes are somewhat ludicrous. It is quite rational not to ignore criticism and not to presume to know how forceful or valid a new piece of criticism may be: continuing a debate while keeping open minds should reveal that. Likewise, ignoring all validly criticized theories is unwise, even though it is a popular attitude: there is no need to ignore either a validly criticized theory or any criticism that turns out to be invalid; at times it is wise to think once on the supposition that a given piece of criticism is devastating and once, to the contrary, that is satisfactorily answerable. What is important overall is to notice that contrary to some strongly felt intuition, valid criticism is progress, and finding that some seemingly strong criticism is weak is a disappointment. But, above all, it deserves notice that the search for the truth invites criticism and learning from experience is learning from criticism. The only significant question in this context is, when do we give up a line of investigation as utterly hopeless?

This, of course, is largely a matter of time and opportunity: when time is short and decision is urgently invited, then we act on less deliberation than otherwise. Moreover, this is why normal society sets time for long-term considerations, for the study of the large trends. And the most important trends are intellectual.

The dominance of rationalism or scientism in Western culture is what gave rise to the greatest and most exciting evolutions of ideas in human history, yet they are also responsible for some very severe backlashes and global risks. The rationalist tradition took it for granted that science will benefit humanity although it may involve some risks. Already Boyle, one of the founders of the scientific tradition as we know it, was aware of this, and even in detail. For example, he bought a patent on some weaponry so as to prevent its use—quite unsuccessfully, of course. But there was no doubt expressed in the scientific tradition that on the whole science will benefit humanity, and that whatever the ills of current science, it will overcome them. The criticism of science, thus, was always viewed as local: it was a global beneficial trend for sure.

The earliest criticism of science as a whole is the doctrine of alienation. It was launched from the viewpoint of the romantic philosophy that saw science as limited to the non-human world (the I-it language, as Martin Buber called it) and viewed the human sphere as different (the I-you relation) and ascribed to that sphere a different methodology.[2] As Uriel Tal has described it, leading nineteenth-century German scientists were of the romantic opinion and recommended that Jews convert to some Christian denomination for political reasons, regardless of their loss of all interest in anything religious.[3] This led to profound conservatism and pessimism in matters social, though not in matters scientific, as can be illustrated with the case of Planck. For reasons all too obvious to articulate, this position is inconsistent; in particular, it prevented Planck from combatting Nazism effectively. The optimist global picture altered with two events: Auschwitz and Hiroshima.

NOTES

1. Thomas S. Kuhn, *The Structure of Scientific Revolutions*, 1962.
2. Martin Buber, *I and Thou*, translated by Ronald Gregor Smith, 2013 (1923).
3. Uriel Tal, *Christians and Jews in Germany: Religion, Politics and Ideology in the Second Reich 1870–1914*, translated by Noah Jonathan Jacobs, 1975.

Chapter 14

Democracy as Optimism

Auschwitz and Hiroshima, these two mega-traumas of the modern world, fortunately do not signify the end of optimism; they signify the end of the naïve optimism of modernism that characterizes the age that has preceded it, the age that ended with two world wars. What characterizes modernity then is not only its political expression as the liberal nation state, but also the disappointment from it that Auschwitz and Hiroshima have come to symbolize. Now this disappointment comes from high expectations, as all disappointments do. It is the responses to disappointments that vary. Significantly, the list of possible responses that disappointment leads to has two significant default options: the extreme ones. One response is giving up the disappointing project altogether. The other response is the minimal reduction of the expectations in question that the disappointment in question imposes. The proposal to give up the project in question, namely, the disappointment in democracy, came soon after the failure of the French Revolution and the democracy that came in its wake. This failure found its expression at once, in the French Terror and the following Napoleonic Wars; the reaction to it was the reactionary movement of the nineteenth century. It was G. W. F. Hegel (1770–1831) who headed it. He rightly viewed the Enlightenment Movement as the cause of the French Revolution, and that made him oppose that movement. He rightly viewed its teaching individualism—the autonomy of the individual as such—as its vey heart, and so he developed the first modern anti-individualist social philosophy. This clinched the dogma that was the consensus of all social philosophy, the erroneous view that individualism and collectivism are mutually exclusive, that by definition the assumption that both individual and society exist is inconsistent. The reason for this is that the assumption that both exist independently of each other invites the assumption that a conflict between individual and society is irreconcilable. Now this assumption is traditional: it is the doctrine of the evil in Man or of Original Sin. Yet the Enlightenment Movement arose out of the rebellion against this doctrine, out of the assumption that Man is good. This is the exaggeration that the French Terror and the Napoleonic Wars refuted. This is the foundation of the Reaction.

The Reaction included rejection of the very values of the Revolution, the values that found their expression in the philosophy of the Enlightenment Movement, often considered as rationalism even though it was a version of rationalism that has turned out to be too optimistic. Traditional rationalism is clear, and so is traditional irrationalism, the rejection of the very idea that rationally constructed politics is the best (and that Tradition is the better guide than Reason).

The case of the unlucky is complicated. Admittedly, although Hegel ignored the presence of alienation in traditional communities, he was right in suggesting that, unlike civil societies, communities have means for combatting alienation. These may solve some problems, but aggregate others. As traditional community imposes sociability on its members, it is unwise to reinstate it unchecked (as Hegel has suggested). And, of course, he was in error when he dismissed the equality that civil society grants the individual. For, civil society recognizes the choice of association. Nevertheless, his criticism is valid. To answer it civil society needs a better education system, one that trains the individual in ways of developing one social environment and it needs traditions that enhance this. Modern education should consist of education for autonomy, and the autonomous should be able to decide whether to request help or not; instead, it educates individuals to avoid such requests as expressions of weakness and thus as lack of autonomy. This is a serious error: the inability to request help is a severe form of disability. Admittedly, education in traditional society is for dependence on others; this does not mean that education in modern society should be for full independence of the individual. Indeed, this is the hallmark of romanticism. The romantic philosophy that characterizes the nineteenth-century reactionary movement made allowance for the autonomy of the individual, but while stressing the fact that most of the ordinary members of extant society are unable to be autonomous. They developed the theory of the hero: the individual of the enlightenment movement is exceptional. He goes to the desert for forty days and returns to a society that rightly ostracizes him. He will die lonely or conquer his society. This will render his ejected doctrine acceptable and practicable. Thus, truth is relative.

That putative truth is not final is a fact; ignoring it is the permanent source of needless disappointment. And so, it is useful to speak of relative truth. What the irrationalist philosophers say is not only that truth is relative: they say that the absolute truth does not exist. This is dreadful. Admittedly, we do not know the absolute truth. Admittedly, we do not know if human language is rich enough to articulate it, yet truth is of an assertion within a given language (as Alfred Tarski has argued). What is important to notice is that putative truth is ever so often false; and falsehood is indeed absolute.[1]

The idea of civil society clashed severely with the idea of traditional community: the ideal community is of interlocking individuals, of constant interdependence; that of the civil society is that of Robinson Crusoe: the (almost) totally independent individual. Both are ideal types, and both are objectionable, and in principle so. And that, to repeat, is the recognition that conflict is unavoidable. It is an odd fact that the first social philosopher who said it was Georg Simmel, who died in 1920, and whose ideas were discarded or ignored until the late twentieth century! He did not deny that conflicts can be contained; on the contrary, this is what he recommended. But he stressed that social philosophy before him was utopian in its avoidance of recognition of the utopian character of frictionless society.

Modern life needs both traditional community and civil society. Indeed, in the better societies in the modern world this is the current situation, and it does, indeed, reduce its problem and conflicts to manageable sizes. More it is impossible to say in general, since troubles appear in different ways in different communities and societies, and so they invite detailed factual studies and political proposals for change, proposals to be tested. To be better prepared for this, parliaments should adopt some constitutional laws requiring all legislation to be temporary and subject to reassessment prior to making them quasi-permanent.

There is a uniformity of fashionable opinions here that the world has paid a high price for. Prior to the French Revolution, the dominant fashion was that of the Enlightenment Movement. The perceptions of this movement were exaggerated in the optimist direction. The optimism of the Enlightenment Movement was striking. Give up your preconceived notions, its members suggested, and listen to the advice that rests on science. And then science will help cure all the maladies of the world. These begin with poverty that invites eradication by the use of science. Indeed, in this the success of applied science is most impressive. Whereas in traditional society most labor by far was invested in agriculture, in the modern world less than one twentieth of it is, and largely due to the ingenuity and perseverance of one person, chemist Justus von Liebig (1803–1873). Nevertheless, more than half of humanity go to bed hungry. This failure, however, is political, no longer due to the poverty of extant resources. The second ill that the Enlightenment Movement was supposed to cure through the use of reason was war. Now the elimination of war should be easier (in principle) than the eradication of poverty, since it is a human product rather than a given. And there is no reason for war other than foolishness and mistrust. For, as Ben Franklin has noticed, there never was a good war or a bad peace. Yet, the elimination of war by the institution of international forums for conflict resolution has failed miserably. The idea led to the formation of the League of Nations that sadly failed to prevent World War II and to the United Nations Organization that granted sponsorship to the Korean War. The League of Nations was a failure, analysts said, because it had no power. And so, during World War II analysts prepared the ground for the formation of the United Nations Organization that had force—not an army of its own but an ability to mobilize armies of member nations. The planners did not build sufficient means to prevent the misuse of the possibility of misuse of the United Nations Organization. To be precise, they did consider this possibility and created means to block it, but in vain. The saddest aspect of the situation is the devastating effect of disappointment: after two failures, the need to try a third time is too often met with incapacitating skepticism. It is here that optimism is so much in need. The need for an effective international organization to settle conflicts is higher due to the new ability of humanity to destroy itself.

This possibility is new: it occurred for the first time after World War II. It was the birth of global politics—not to be confused with international politics. International politics is as old as nations; global politics is a new discipline; it concerns problems that cannot find solutions in international politics. A global problem is one that cannot find local solutions, that may possibly have solutions, but, if so, then only in the global dimension. There are four of these, the four P's: *P*overty, *P*ollution, *P*opulation-explosion, and the *P*roliferation of nuclear weapons.

Optimism is never to lose hope—regardless of circumstances. The Marquis de Condorcet is a good example for that. He was a victim of the French Terror, yet while he hid from the wrath of the authorities, he wrote in his hiding place that the success of the future of humanity is assured. His proof was simple and short: everyone will emulate any successful innovation, so that these must accumulate. That destructive innovations would lead to two bloody world wars and endanger the very future of the human race was beyond his imagination. The two world wars were indeed a novelty: they were total: they involved civilians as much as they involved soldiers; more civilians than soldiers died in each of the two world wars. Russell deemed them as one prolonged war; yet when the first one started, people expected that that it would end within a few months. For, its aim was merely to determine the division of the globe between empires. Imperialism then is the force behind these disasters. Even Russell had supported it for a year or two, he reported late in life. What was its allure then? No answer. Perhaps the reason was paradoxically the very rise of modern democracy. It was Spinoza who said, since no governments can rule by relying on brute force alone, all regimes are in principle democratic to this or that extent. And so, he concluded from this observation, the central political question is, how much democracy is desirable? The answer he gave is, the more the better.

Now, before World War I, Great Britain was democratic in principle. In fact, full universal suffrage did not exist there prior to the end of World War II. What was much worse, is that the democratically chosen rulers were aristocrats. This changed slowly. First some leading scientists were from "lower walks of life," as the Britannica said of Sir Humphry Davy (1778–1829). His heir was Faraday (1794–1867) who came from the lowest walk of life. In other words, science offered a new means for social climbing. Then George Jacob Holyoake (1817–1906) invented the cooperative movement. And then, around 1900, the labor party had as a member of the British parliament Keir Hardie, son of a of coal miner. This was found shocking. The democratizing process began to threaten the aristocrats who had managed to stay in power until then. As the higher classes lost their position as the national political leadership, they sought a social task elsewhere. They found it in the white man's burden, which is the politically correct name for imperialism.

Imperialism was hardly eve new, of course. Ancient empires extracted loyalty and taxes from their subjects. The Middle Ages had no imperialist aspirations. Islam was the political idea of uniting the whole civilized world under one peaceful religion. This idea was wild but not beyond success. As it happened it failed with the first failure of the expanding Muslim forces: they never managed to go north of Iberia, and their expansion eastward had no influence on Europe that remained Christian and had no imperialist ambitions. The Dutch, British, and French (later also Prussian) empires were not due to any aspirations in their homelands. These grew around 1900 and led to two world wars. Moreover, when expansion did appear, it was in north Africa, where the French wished to settle rather than rule the native population. This, to repeat, happened around 1900. Nevertheless, the much earlier popular imperialist aspirations must have played some role in the historical development. Jules Verne's early (1865–1867) novel, *In Search of the Castaways*, shows how strong and how vague these aspirations were. This is of a particular importance in discussions of aspirations for democracy since imperialism was always a diversion of discussions about democracy to other channels. The most significant of these diversions was the question, very popular in the 1930s: Do you side with communism or with fascism? In this question the idea hides that democracy is a dead option.

Strange. Democracy appears repeatedly both as attractive and as impossible. Its attraction rests on its liberalism: it is in the liberal preference for anarchy, literally no rule, and the recognition of its impossibility. The Bible describes the period prior to the Kingdoms of Saul and David as anarchy, and as ideal. Leading modern Bible scholar Martin Buber was an anarchist and viewed that book as describing an ideal. The question that an anarchist has to answer is, how is anarchy with no chaos possible? The received opinion is that it is not. Indeed, in common parlance the word "anarchy" is used to denote political chaos. This, indeed, is the bone of contention. When asked to describe anarchy, the anarchist says, imagine the current regime as it is but with no police and no army. There will be conflicts, and there will be law courts, but no coercion. People will willingly do what they do now only under duress. If this is their disposition, why do they not behave in accord with it even here and now? Because we possess armed forces. Hence, crime is the outcome not of what criminologists tell us it is but because we have armed forces. This sounds odd. Yet there is a model for it: usually police forces bear arms, yet the vast majority of the British police force do not. In other countries the police say that they have to bear arms, since the population is not as civilized as the British; anarchists put it the other way round: populations are not as civilized as the British just because their police forces are armed.

Whatever is the case, clearly, disagreement about it is reasonable, but about the preferability of effective unarmed police or armed police is not. And this is the charm of democracy, as is manifest in the preferability of a peaceful society over a lawless one. (It is thus important to see that desirable anarchy is not lawless but free of all bullying.) That this is the charm of democracy can be seen from the view of minimal democracy: at the very least, say Russell and Popper, democracy limits the arbitrary use of force of the people in power: democracy can dethrone them with no revolution and no bloodshed. The advantage of this idea is that it is minimalist. As a matter of course, Russell and Popper want much more of democracy, but at the very least they expect that it will defend citizens from too much arbitrary intimidation of ordinary citizens.

Under what conditions may a society try to disarm its police force? This depends on what we expect of our society, and again, to avoid utopianism we should aim at the minimum. There are, in historical fact, two candidates for this status, (liberal) civil society and (conservative) community. The most significant difference between them concerns equality. This is why hyper-conservative Hegel derided it as merely formal.

The two world wars had much less to do with that than with the trenches. Wars were traditionally fought between two military forces facing each other (unless they were guerilla wars, namely when the forces were too unequal to face each other squarely). Some important wars were decided by one decisive battle each. Even longer wars often comprised separate battles each fought on its own. And then World War I lasted four bloody years and terminated in one battle—one in which tanks replaced the cavalry. The result was devastating and took a generation to recover from. No moral from this was to be learned. The recovery followed by the second round that was even bloodier, although trenches played in it much less significantly than in its predecessor. The reason no lesson was learned from World War I to prevent World War II is the humiliation of Germany in the peace accord that followed World War I. Even civilized people like Planck the revered father of quantum mechanics felt that Germany should do something to recover its lost honor! But then Planck was no democrat. It was the democratization of devastated Germany that prevented World War III no less than its shame in the face of the Shoah that they had initiated. In that case there was an experiment to test this: Germany was divided, with East Germany clearly undemocratic and West Germany in dubious democracy—dubious because it was imposed by the victorious former enemies.

The obvious aspect of the situation eludes many students of social affairs, sociologists, politologists, economists and criminologists alike. It is that democracy is never perfect so that any fairly reasonable criticism of it is likely to be valid, so that democracy should take the admission of criticism as the default option rather than the parrying it. Often critics of democracy launch their criticism under the illusion that valid criticism of democracy should rationally lead to its abandonment. Yet both tyranny and democracy can survive valid criticism and the consequent admission of it. The best criticism of democracy ever is the one that Plato has launched more than wo millennia ago: democracy leads to anarchy and anarchy invites unscrupulous individuals to try to rule it tyrannically. The alternative to democracy that Plato offered is the rule of a philosopher-king—meaning Plato himself. As it turns out, often the success in crowning a philosopher-king ends up in a disaster: the ruler turns out then to be either inept or ruthless. Another option is possible, of course: democratization. As it happens, most efforts at democratization have ended up in failure. This raises an interesting question: how then does the currently more successful part of the world comprise liberal democracies?

Before answering this intriguing question, it behooves us to admit that many commentators, among them serious, internationally recognized scholars contest the supposition behind this question: they deem liberal democracy an obvious failure. They do have many powerful arguments on their side: all known democracies are deficient. This is obvious, even *a priori* so: there is no utopia. Nevertheless, it is obvious that they lack a sense of proportion. Admittedly, we live in a world whose future is in doubt; yet the world is free of the slavery and the piracy that were ever present in the more traditional world, and many countries are at peace with their neighbors that is not threatened by an imbalance of power. Moreover, although no country has succeeded in eradicating poverty and hunger, in many countries these have ceased being the most urgent issues on their national agenda.

To the question, then: how does the currently more successful part of the world comprise liberal democracies? Whence the success of the democratic experiment? To repeat, since democracy tends to invite anarchy, it usually ends with failure; whence, then, its current success? The answer is, paradoxically, that politics interest less people today then in times past. And that even those interested in it are less passionate. Why is that? Because economic prosperity makes politics less attractive. Proof: currently all dictatorships succeed in poor countries; the rich countries are democratic and democracies are rich. This proof is faulty: both China and Russia are rich, they are kleptocracies and autocracies and relatively stable. What is the cause and what is the effect here is hard to judge, and the true answer is apparently that it was a bootstrap operation. This explains why it is so hard to export democracy. After World War II the United Nations Organization was in charge of dismantling empires and the liberated countries were created democracies, but to no avail. Thus, the Belgian Congo which charitable Belgian political scientists offered a whole new constitution was democratic only for a few hours. This suggests the need for a mechanism for step-by-step economic and political progress. Optimism suggests that such a process is successfully conceivable and its outcome successfully implementable.

NOTE

1. Karl Popper, *The Logic of Scientific Discovery*, §84, 1959.

Conclusion

Autonomous People in Free Society

Traditional philosophy was supposed to navigate successfully between dogmatism and skepticism. This brief and powerful adage invites expansion, as it is too often misunderstood. That dogmatism is undesirable goes without saying: the dogmatic is intellectually stuck. That skepticism is also undesirable is less obvious. Indeed, it may be very desirable, at least as a response to dogmatism: it behooves us to doubt the dogmas that we face. Hence, it is only skepticism as such that may be undesirable, skepticism as applied indiscriminately to all ideas. Why? To find this it is necessary to go back to dogmatism: what is the cure of it?

All this is traditional, and so is the cure proposed against it, the one that is quite reasonable and that nevertheless moved philosophical discourse off its trail. The opposite of a dogma as such is logic as such: whatever is logical surely is no dogma. What then is logical? Classical logic offers a clear answer: valid inferences are logical. Yet the truth of a conclusion to a valid inference is guaranteed if and only if its premises are. When are premises assuredly true? When they are proven. What is proof?

All this is traditional, except for the last three words: what is a proof? Oddly, although tradition requires of thinkers to prove their ideas, for millennia this question was neglected. It was Kant who asked, what kind of assertion is provable? The reason for his inquiry is this:

Newtonian mechanics was deemed empirically proven, even though Newton himself said clearly that no empirical theory can be fully proven. It was Hume who proved this: empirical proof is impossible. Kant asked then, how can we prove it nonetheless? He said, as standard logic as a proof theory is powerless, we need another logic. He called it transcendental logic. To make room for it he first developed proof theory and limited standard logic or analytic logic as he called it and then he launched synthetic logic, the logic of proof of synthetic statements. For this his candidates were geometry and physics. This led Einstein to his wonderful claim: geometry is either proven or about the world but not both. He dared supersede Newton. The famous equation of mass and energy was not new. But it applied only to electromagnetic fields. There were two systems in physics, mechanical and electromagnetic; efforts were made to force fields into the mechanical mold. He turned things around: he forced mechanics into the electromagnetic mold. This is the view of matter as compact energy. It looks disturbing. To put it in traditional Aristotelian terms, Einstein showed that the actual is potential.

Thus, the cat that sits by the hole in the wall, waiting to pounce on the mouse when he comes out, is not separate from the disposition to pounce: it is that disposition. The disposition for what? For pouncing; yet pouncing too is but a disposition. Is there anything other than a disposition, a more tangible something? Now to be tangible is a disposition, a disposition to touch; yet to touch is also to be disposed to something or another.

This showed how much more freedom researchers have if they deny the possibility of poof in science. It allows for revolutions in science—indeed, for progress as revolutionary. What then is the status of old theories? They are special cases and first approximations.

All this should hold *a fortiori* in the human sphere, social and psychological.

Ever since Descartes, who postulated two substances, matter and mind, efforts were made to render social science a part of psychology. Opposite Descartes were the monists who postulated one substance, matter or mind. The view of mind as substance assured its immortality; the view of mind as inherently material does the opposite and so it seemed religiously heretical. All this is outdated: Einstein's 1905 revolution does away with substance. The same holds, incidentally, for Russell's revolution in logic, also in 1905: substance in the sense of Thales of old is unchangeable, and so there can be no more than one substance in the world, he observed, as already Spinoza understood. Moreover, the substance theory does not allow for change. This led Parmenides to the idea that motion is illusory, and Heraclitus said that there is no substance. Democritus offered a compromise: atoms are unchangeable, but they move in empty space: he assumed two substances, atoms and the void. This did more than solve a central problem that looked insoluble: he showed that ingenuity allows for utterly new solutions, for the possibility that the abandonment of hope may be rash even if it looks desperate.

This looks decidedly good news. Some, however, are unhappy about it: the result of increased freedom is added problems of choice. These may be overwhelming! Most members of traditional society are unaware of all this: they live in stable societies with fixed routines, most of which pose no problem and other that pose simple decision problems with obvious possible solutions and easy choices between them. Many educated people in Western countries are quite capable to decide to be autonomous yet while never having faced the need to decide on the matter. Individuals of this sort may face the problem—whether to be autonomous or not—as for example by being informed of the very possibility. They then may decide to be autonomous. Such a decision is reversible, but only for a brief period of grace. Afterwards, they lose the ability to reverse to heteronomy: they are unable to accept any authority unconditionally. Sartre lamented this inability as the imposition of independence. He was right, although this is hardly a proper subject for any lament.

Though erroneous, the lament is reasonable. Autonomy is good when limited to leaders and prospective leaders, Plato observed, but a cause for social instability when prevalent. This predicament is general: autonomous or not, individuals who satisfy their own needs and wishes do not contribute to society and are thus parasitic and thus antisocial. Hence, the conflict between individuals and their societies is inevitable. And so, individuals must yield, since society can live without any individual, but no individual can live without society.

This comprises a great incentive to develop social science and to show that sociology is but the psychology of a multitude. This idea has been refuted.[1] Yet the fact remains: most of the standard applications of the idea that individuals have to yield to society have turned out to be mistaken since much of the sacrifice that individuals gave up for the sake of society turned out to be useless. What society benefits from is what autonomous individuals did, in total disregard of the needs of their societies. As Adam Smith has noted, individuals contribute most to their societies when they act in self-interest. His intent was to encourage innovation. Although his theory came to boost competition, he approved of patents that exclude competition as he considered the profit from a patent just and socially beneficial. This seems common sense, yet it assumes two questionable assumptions. One is eudaimonia: action in enlightened self-interest is always just. The other is that enlightened people have to act in enlightened self-interest. Already Hume noticed that self-sacrifice refutes this. He brushed this refutation aside. This may be reasonable in view of the scarcity of the need for such conduct. Yet the overriding assumption of all versions of eudaemonia is that people are autonomous. The strongest valid criticism that the nineteenth-century Reaction launched against the philosophy of the Enlightenment is the claim that most people are not autonomous and are unable to thrive as autonomous beings.

This critique is important and valid, but it is not fair. Fairness should draw attention to the fact that autonomy requires instigation and encouragement. It is no accident that it is rare in traditional societies but common in modern democracies. It is, indeed, what keeps democracies stable.

We want our offspring to be as free as possible to choose their own lifestyles. This does not work for autonomy since education for it blocks the way to its abandonment. Parents must choose whether to educate their offspring as autonomous or heteronomous, and those who are aware of it take it for granted that they should try to educate their offspring to grow as autonomous.

This is reasonable in democratic societies, and this makes parents in democracies take their democracy and their autonomy for granted—as one. As long as one lives in a self-governing democratic society, one may take it for granted. Yet when one's society is politically not self-governing or otherwise in need of reform, things turn out to be not that obvious. This is not merely an abstract exercise. Its application to the improvement of working life raised it as a serious problem. The industrial revolution created a new kind of employment, factory work, that is notoriously intolerable, as shown in literature and movies. Taking autonomy as the right option in education is assuming democracy as its background. This we often take for granted, although it is food for thought.

NOTE

1. Karl Popper, *The Open Society and Its Enemies*, Ch. 14, 1945.

Bibliography

Agassi, Joseph. "Leibniz's Place in the History of Physics." *Journal of the History of Ideas* 30(3) (1969): 331–344.

———. *Towards a Rational Philosophical Anthropology*. The Hague: Martinus Nijhoff, 1977.

———. *Science and Society: Studies in the Sociology of Science*. Dordrecht: Reidel, 1981.

———. "The Unity of Hume's Thought." *Hume Studies* 1 (1985): 87–109.

———. "Hegel's Scientific Mythopoiesis in Historical Perspective." In *Studies in the Philosophy of J. N. Findlay*, 445–458. Edited by R. S. Cohen, R. M. Martin, and Merold Westphal. Albany, NY: State University of New York Press, 1985.

———. *Radiation Theory and the Quantum Revolution*. Basel: Birkhäuser, 1993.

———. "Reason within the Limits of Religion Alone: The Case of Maimonides." In *Relativism and Beyond*, 125–177. Edited by Yoav Ariel, Shlomo Biderman, and Ornan Rotem. Leiden: Brill, 1998.

———. *Science and Its History: A Reassessment of the Historiography of Science*. Cham: Springer, 2008.

———. *Ludwig Wittgenstein's Philosophical Investigations an Attempt at a Critical Rationalist Appraisal*. Cham: Springer, 2018.

Agassi, Joseph, and Ian C. Jarvie, eds. *Rationality: The Critical View*. Dordrecht: Nijhoff, 1987.

Al-Farabi. *Philosophy of Plato and Aristotle*. Ithaca, NY: Cornell University Press, 1969.

Aquinas, Thomas. *Summa Theologiae*. Edited by David John Bourke. Cambridge: Cambridge University Press, 2006.

Bacon, Francis. *Novum Organum* "New Method." Bottom of the Hill, 2012 (1620).

Banton, Michael. *Roles: An Introduction to the Study of Social Relations*. London: Tavistock Publications, 1965.

Bloom, Allan David. *The Closing of the American Mind*. New York, NY: Simon and Schuster, 1987.

Buber, Martin. *I and Thou*. Translated by Ronald Gregor Smith. London: Bloomsbury Academic, 2013 (1923).

Butler, Samuel. *Erewhon*. London: Trubner & Co., 1872.

Cohen, Robert S., Richard M. Martin, and Merold Westphal, eds. *Studies in the Philosophy of J. N. Findlay*. Albany, NY: State University of New York Press, 1985.

Copernicus, Nicolaus. *On the Revolutions of the Heavenly Spheres*. Translated by Charles Glenn Wallis. Amherst, NY: Prometheus Books, 1995 (1543).

Coser, Rose Laub. "Authority and Decision-Making in a Hospital: A Comparative Analysis." In *Medical Men and Their Work*, edited by Eliot Freidson and Judith Lorber. New York, NY: Aldine Publishing Company, 1972.

Descartes, René. *Œuvres de Descartes*. Edited by Charles Adam and Paul Tannery. Paris: J. Vrin, 1996.

Fanon, Frantz. *The Wretched of the Earth*. New York, NY: Grove Press, 2007 (1961).

Feyerabend, Paul K. *Knowledge Without Foundations*. Oberlin College, Open Library, 1961.

Freud, Sigmund. *Psychopathology of Everyday Life*. New Delhi: General Press, 2018 (1901).

———. *Civilization and Its Discontents*. Translated by James Strachey. New York, NY: W.W. Norton, 2010 (1930).

Gabbay, Dov M., and John Woods. *A Practical Logic of Cognitive Systems*. Amsterdam: North Holland, 2003.

Galilei, Galileo. *Dialogue Concerning the Two Chief World Systems: Ptolemaic and Copernican*, translated and with revised notes by Stillman Drake. New York, NY: The Modern Library, 2001 (1632).

Gellner, Ernest. *Cause and Meaning in the Social Sciences*. New York, NY: Routledge, 1973.

Gilbert, Martin. *The First World War: A Complete History*. New York, NY: Henry and Holt Company, 1994.

Gilbert, W. S. *HMS Pinafore*. Edited by Percy Young. New York, NY: Broude Brothers Limited, 2003.

Herodotus. *The Histories*. Harmondsworth: Penguin, 1977.

Hume, David. "Of the Rise and Progress of the Arts and Sciences." In *Essays Moral, Political and Literary*. New York, NY: Wallachia Publishers, 2015 (1776).

Huxley, Aldous. "The New Comfort." *Vanity Fair*, April 1927. https://archive.vanityfair.com/article/1927/4/the-new-comfort.

———. *Brave New World*. London: Chatto & Windus, 1932.

———. *After Many a Summer*. New York, NY: Harper & Bros., 1939.

Idel, Moshe. "Johannes Reuchlin: Kabbalah, Pythagorean Philosophy and Modern Scholarship." In *Representing God*, 123–148. Leiden: Brill, 2014.

Jackson, Bernard S., ed. *Studies in Jewish Legal History in Honour of David Daube*. London: Jewish Chronicle Publications, 1974.

Jarvie, Ian C. *The Revolution in Anthropology*. New York, NY: Routledge, 1964.

———. *Philosophy of the Film: Epistemology, Ontology, Aesthetics*. New York, NY: Routledge, 2003.

Kant, Immanuel. *Critique of Pure Reason*. Edited by Paul Guyer and Allen W. Wood. Cambridge: Cambridge University Press, 1998.

Kipling, Rudyard R. "The Man Who Would Be King." In *Under the Deodards*. Prayagraj, India: A.H. Wheeler, 1888.

Kroeber, Alfred L. *Checklist of Civilizations and Culture*. New York, NY: Wenner-Gren Foundation for Anthropological Research, 1962.

Kuhn, Thomas S. *The Structure of Scientific Revolutions*. Chicago, IL: The University of Chicago Press, 1962.

———. "Replies to My Critics." In *Criticism and the Growth of Knowledge*. Edited by Imre Lakatos and Alan Musgrave. Cambridge: Cambridge University Press, 1970.

Lakatos, Imre. *Proofs and Refutations*. Cambridge: Cambridge University Press, 1976.

Leavis, F. R., and Stefan Collini. *Two Cultures? The Significance of C.P. Snow*. New York, NY: Cambridge University Press, 2013 (1962).

Lessing, Gotthold E. *Nathan the Wise: a Philosophical Drama*. Translated by Rudolf Erich Raspe. London: J. Fielding, 1781 (1779).

Locke, John. *An Essay Concerning Human Understanding*. Edited by Peter H. Nidditch. Oxford: Clarendon Press, 1975.

Machiavelli, Niccolò. *The Art of War*. Edited by Neal Wood. Cambridge, MA: Da Capo Press, 1965 (1621).

McGuinness, Brian. *Moritz Schlick*. Dordrecht: Reidel, 1985.

Mead, Margaret. *Coming of Age in Samoa: A Psychological Study of Primitive Youth for Western Civilization*. Gloucester, MA: P. Smith, 1978 (1932).

Morgan, Lewis Henry. *Ancient Society or Researches in the Lines of Human Progress from Savagery Through Barbarism to Civilization*. Chicago, IL: Charles H. Kerr & Company, 1907 (1877).

Needham, Joseph. *Science and Civilization in China*. Cambridge: Cambridge University Press, 1954–present.

Nietzsche, Friedrich. *Sämtliche Werke*. Edited by Giorgio Colli and Mazzino Montinari. München: Deutscher Taschenbuch Verlag, 1980.

Orwell, George. *A Clergyman's Daughter*. London: Victor Gollancz, 1935.

Pestalozzi, Johann Heinrich. *Enquiries into the Course of Nature in the Development of the Human Race*, 1797.

———. *Leonard and Gertrude*. Translated by Eva Channing. Boston, MA: DC Heath and Co., 1891.

Polanyi, Michael. *Science, Faith and Society*. Chicago, IL: University of Chicago Press, 1964.

Popkin, Richard H. "The Pre-Adamite Theory in the Renaissance." In *Philosophy and Humanism: Renaissance Essays in Honor of Paul Oskar Kristeller*, 50–69. Edited by Edward P. Mahoney. New York, NY: Columbia University Press, 1976.

Plato. *Complete Works*. Edited by John M. Cooper and D. S. Hutchinson. London: Hackett Publishing, 1997.

Popper, Karl R. *Logik Der Forschung*. Vienna, Austria: Verlag Von Julius Springer, 1935.

———. *The Open Society and Its Enemies*, 2 vols. London: Routledge, 1945.

———. *The Logic of Scientific Discovery*. London: Hutchinson & Co., 1959.

———. "Towards a Rational Theory of Tradition." In *The Rationalist Annual 1949*. London: Watts & Co., 1949.

Porta, Giambattista della. *Magia Naturalis*, 1558. English version, *Natural Magick*. London: Thomas Young and Samuel Speed, 1659.
Proust, Marcel. *À la recherche du temps perdu*. Paris: Gallimard, 1975.
Rousseau, Jean-Jacques. *Emile, or On Education*. Translated by Allan Bloom. New York, NY: Basic Books, 1979 (1762).
Russell, Bertrand. "On Denoting." *Mind* 14 (1905): 479–493.
———. *Problems of Philosophy*. London: Williams and Norgate, 1912.
———. *Marriage and Morals*. London: Allen and Unwin, 1929.
———. "The Cult of 'Common Usage.'" *The British Journal for the Philosophy of Science*, 3(12) (February 1953): 303–307.
———."Zahatopolk." In *Nightmares of Eminent Persons and Other Stories*, 67–109. London: The Bodley Head, 1954.
Said, Edward W. *Orientalism*. New York, NY: Pantheon Books, 1978.
Sartre, Jean-Paul. *The Words*. New York, NY: Fawcett Crest Books, 1963.
Shankman, Paul. *The Trashing of Margaret Mead: Anatomy of an Anthropological Controversy*. Madison: University of Wisconsin Press, 2009.
Shaw, Bernard. "The Doctor's Dilemma." In *The Doctor's Dilemma, Getting Married, and The Shewing-up of Blanco Posnet*. New York, NY: Brentano's, 1911 (1904).
Spinoza, Benedictus de. *The Collected Works of Spinoza*. Edited by Edwin Curley, vols. 1 and 2. Princeton, NJ: Princeton University Press, 1985; 2016.
Tal, Uriel. *Christians and Jews in Germany: Religion, Politics and Ideology in the Second Reich 1870–1914*. Translated by Noah Jonathan Jacobs. Ithaca, NY: Cornell University Press, 1975.
Tönnies, Ferdinand. *Gemeinschaft Und Gesellschaft*. Berlin: De Gruyter, 2019 (1887).
Verne, Jules. *In Search of the Castaways*. Paris: Hetzel, 1865–1867.
Weber, Max. *The Protestant Ethic and the Spirit of Capitalism* (1905) *and Other Writings*. Edited, translated, and with an introduction by Peter R. Baehr and Gordon C. Wells. New York, NY: Penguin, 2002.
Wells, H. G. *The New Machiavelli*. London: The Bodley Head, 1911.
———. *The Island of Doctor Moreau*: The Original 1896 Edition, Suzeteo Enterprises, Open Library, 2020.
Wolfson, Harry A. *The Philosophy of Spinoza: Unfolding the Latent Process of His Reasoning*, 2 vols. Cambridge, MA: Harvard University Press, 1934.
———. *Philo: Foundations of Religious Philosophy in Judaism, Christianity, and Islam*, 2 vols. Cambridge, MA: Harvard University Press, 1947.

Index

About the Author

Joseph Agassi, FRSC, studied in a rabbinical school, in the Hebrew University of Jerusalem (MSc, physics) and in the University of London, where he was assistant to Karl Popper (PhD, general science: logic and scientific method). He was a parachute instructor in the Israeli Defense Force. He is professor emeritus in Tel Aviv University and in York University, Toronto, and *membro onorario del Corpo Accademico dell'Universita degli Studi "Gabriele d'Annunzio,"* Chieti-Pescara, Italia.

He has edited about ten books, written about twenty, and made about 600 contributions to the learned press. Among them are *Science in Flux*; *Science and Society*; *Science and Culture*; *Science and Its History*; *The Siblinghood of Humanity: Introduction to Philosophy*; *Towards a Rational Philosophical Anthropology*; *Liberal Nationalism for Israel: Towards an Israeli National Identity*; *The Hazard Called Education*; and *Ludwig Wittgenstein's Philosophical Investigations: An Attempt at a Critical Rationalist Appraisal*. He coauthored, with Yehuda Fried, *Paranoia: A Study in Diagnosis* and *Psychiatry as Medicine*; with Nathaniel Laor, *Diagnosis: Philosophical and Medical Perspectives*; with Abraham Meidan, *Philosophy from a Skeptical Perspective*; with Ian C. Jarvie, *A Critical Rationalist Aesthetics*; and, with David Harel, *Malevich: The Lost Paintings*.

www.ingramcontent.com/pod-product-compliance
Ingram Content Group UK Ltd.
Pitfield, Milton Keynes, MK11 3LW, UK
UKHW012118150926
475560UK00002B/171

9 781793 651730